Trade, Climate Change and Sustainable Development

Key Issues for Small States, Least Developed Countries and Vulnerable Economies

Edited by

Dr Moustapha Kamal Gueye, Malena Sell and Janet Strachan

COMMONWEALTH SECRETARIAT

ICTSD
International Centre for Trade and Sustainable Development

Published by
Commonwealth Secretariat
Marlborough House
Pall Mall
London SW1Y 5HX
United Kingdom

Published by the Commonwealth Secretariat
Edited by Jane Lanigan, Editors4Change Ltd
Designed by S.J.I. Services, New Delhi
Cover design by Tattersall Hammarling & Silk
Index by Indexing Specialists (UK) Ltd
Printed by Hobbs the Printers Ltd, Totton, Hampshire

Cover photo credit: Aubrey Wade/Panos Pictures

Copies of this publication may be obtained from
The Publications Section
Commonwealth Secretariat
Marlborough House
Pall Mall
London SW1Y 5HX
United Kingdom
Tel: +44 (0)20 7747 6534
Fax: +44 (0)20 7839 9081
Email: publications@commonwealth.int
Web: www.thecommonwealth.org/publications

A catalogue record for this publication is available from the British Library.

ISBN: 978-0-85092-881-5 (paperback)
ISBN: 978-1-84859-000-7 (downloadable e-book)

Contents

List of tables, figures and boxes

Acknowledgements

This publication is the product of a collaboration between the International Centre for Trade and Sustainable Development (ICTSD) and the Commonwealth Secretariat.

Contributors to this volume include experts and analysts from various institutions in developing and developed countries. It was edited by Malena Sell, Dr Moustapha Kamal Gueye, Samantha Fang and Janet Strachan and Graham Banton, under the direction of Ricardo Melendez-Ortiz, Christophe Bellmann and Vasantt Jogoo.

The opinions expressed are those of the authors and do not necessarily reflect the position of ICTSD or the Commonwealth Secretariat.

Preface

The purpose of this publication is to deepen the understanding of policy-makers and other stakeholders of the major issues and challenges that least developed countries (LDCs), small and vulnerable economies (SVEs) and small island developing states (SIDS) face with respect to the interface between trade and climate change. Previous work in this field has focused largely on the concerns of developed countries and large developing countries such as Brazil, India and China. Yet, for smaller economies, the inter-relationship between trade and climate change is likely to have significant development implications.

LDCs, SVEs and SIDS are amongst the countries most vulnerable to climate change. Not only do many of them face amongst the most severe physical impacts from climate change – often in the form of too much water, or too little – but they also have economies that are particularly sensitive to climatic variation, since these are based in the natural environment and dominated by agriculture, fisheries, forestry and tourism. Furthermore, countries in this group are amongst the most open and trade-dependent economies in the world. Many are remote from major markets, being either islands or landlocked countries, and face high transportation costs. They also face deep institutional and human resource capacity constraints in formulating and institutionalising effective pre-emptive and response measures to climate change and climate change policies.

The study that this book presents was carried out under the International Centre for Trade and Sustainable Development's (ICTSD) Global Platform on Climate Change, Trade and Sustainable Energy. It was implemented by the Commonwealth Secretariat in response to the Lake Victoria Commonwealth Climate Change Action Plan, agreed by Commonwealth Heads of Government in Uganda in November 2007.

The study, in draft form, was considered at a Multi-stakeholder Dialogue on Trade and Climate Change: Key Issues for Developing Countries, which was held in Mauritius in September 2008. The meeting was hosted by the Ministry of Finance and Economic Development of Mauritius and involved researchers involved in the project, regional bodies, the private sector and government officials from Africa, the Caribbean and Pacific. There was also strong representation from within Mauritius across a range of sectors. Some of the policy-oriented conclusions drawn out by the stakeholder dialogue are included in this volume. They indicate a rich programme of work that has still to be developed and implemented to address current development concerns of LDCs, SVEs and SIDS in respect to trade and climate change issues.

The papers in this volume comprehensively cover the trade and climate change concerns of LDCs, SVEs and SIDS, looking at the impact of climate change on key trade sectors, including agriculture, fisheries and tourism, and exploring the needs of these most vulnerable countries with respect to transport, and access to clean, efficient and renewable sources of energy for development, and technologies that will support more sustainable forms of production and livelihoods into the future.

The book highlights some of the most immediate concerns of LDCs, SVEs and SIDS in respect to trade and climate change issues, but the implication of the analysis this volume presents is also that there needs to be a transformation of trade policy approaches, at national and international levels, to support a diversification away from trade sectors that are highly vulnerable to climate change, and to better support economic competitiveness, low-carbon development and poverty reduction in the most vulnerable countries in the world.

Abbreviations and acronyms

BATs	Best-available technologies
CDM	Clean development mechanism
CERs	Certified Emissions Reductions
EGS	Environmental goods and services
EPA	Economic Partnership Agreement
EPPs	Environmentally preferable products
ETS	Emissions Trading Scheme
FFV	Fresh fruit and vegetable
FPJ	*Fresh Produce Journal*
GDP	Gross domestic product
GEF	Global Environment Facility
GHG(s)	Greenhouse gas(es)
HS	Harmonized Commodity Description and Coding System
IATA	International Air Transportation Association
ICTSD	International Centre for Trade and Sustainable Development
IEA	International Energy Agency
IIED	International Institute for Environment and Development
IP	Intellectual property
IPCC	Intergovernmental Panel on Climate Change
ISO	International Organization for Standardization
LCA	Life cycle analysis
LDCs	Least developed countries
MEA	Multilateral environmental agreement
MEPS	Minimum energy-performance standards
NGO	Non-governmental organisation
OECD	Organisation for Economic Co-operation and Development

PPMs	Process and production methods
R&D	Research and development
SAs	Sectoral approaches
SDCs	Small developing countries
SIDS	Small island developing states
SPS	Sanitary and phytosanitary
SSA	Sub-Saharan Africa
SVEs	Small and vulnerable economies
REDD	Reducing Emissions from Deforestation and Degradation
TBT	Technical Barriers to Trade
TRIPS	Trade-related Aspects of Intellectual Property Rights
UNCTAD	United Nations Conference on Trade and Development
UNEP	United Nations Environment Programme
UNFCCC	United Nations Framework Convention on Climate Change
WTO	World Trade Organization

SUMMARY

Introduction

The interface between trade and climate change – both from the perspective of mitigation and adaptation – has entered the international high-level policy arena. The interests and concerns of developed and large emerging economies in this area have received significant attention. However, the prospects and perspectives of smaller developing countries – including small and vulnerable economies, least developed countries and small island developing states – remain obscure. Addressing that gap, this publication provides an analysis of trade and climate change concerns from the broad perspective of the economies of least developed countries, small and vulnerable economies and small island developing states, and goes on to examine a range of relevant sectoral and policy concerns in greater depth.

The Climate Change and Trade Challenges of Smaller Developing Countries

The projected impacts of climate change, as described by the Intergovernmental Panel on Climate Change (IPCC) in its Fourth Assessment Report, will be felt all over the planet. The IPCC projects that by 2020, between 75 and 250 million people in Africa will be exposed to increased water stress, while in some countries, yields from rain-fed agriculture could be reduced by up to 50 per cent. In many parts of Asia, freshwater availability, particularly in large river basins, is projected to decrease. Coastal areas with high population density will be greatly affected by increased flooding from the sea and rivers. Changes in precipitation patterns and the disappearance of glaciers are projected to significantly affect water availability for human consumption, agriculture and energy generation. Adapting to climate change similarly poses challenges. In Africa, the cost of adaptation would consume at least 5–10 per cent of GDP by 2080, which could be 5 to 8 per cent higher in arid and semi-arid areas. Most of the developing world will face related economic and social costs in their adaption efforts.

In response to these tremendous challenges, negotiations on a comprehensive new legal instrument to address climate change are underway, with 2009 representing a critical milestone for an agreement. As laid out in the Bali Action Plan for the negotiations, the UN Framework Convention on Climate Change (UNFCCC) Conference of the Parties will meet in December 2009 to finalise a deal. In this context, it is crucial to promote consensus among countries, and to encourage the informed participation of developing countries in particular. One element in the negotiations will be the role that trade plays and ways that trade policy can be harnessed to support climate change mitigation and adaptation.

In addition, following the informal trade ministers' meeting at the sidelines of the UNFCCC Conference of the Parties in Bali in December 2007, a range of trade topics of great political sensitivity and relevance have entered the climate policy arena with force – and these issues need to be addressed. Key topics include the impacts of climate policies on competitiveness in a global market place where different actors take on different obligations. Calls for responses include those related to border measures, i.e. placing carbon taxes of some form on carbon-intensive imports. Other issues include standards and labelling in the area of energy efficiency, the diffusion through trade of climate-friendly technologies, the 'food miles' debate and carbon footprint of global trade more generally, technology transfer, intellectual property issues, biofuels and biofuel certification schemes, as well as the impact of trade on adaptation needs and measures. Many of these policies and measures considered in response to climate change have crucial implications for developing countries, despite their small contribution to global warming and limited capacity to respond.

While much of the debate on trade and climate change has centred on key developed countries (the European Union and United States in particular) as well as the large emerging economies (especially China, Brazil and India) the implications for and effects on smaller developing countries may be significant; however, such effects are poorly understood.

The smaller countries considered in this book include least developed countries (LDCs), small and vulnerable economies (SVEs) and small island developing states (SIDS).

Trade and climate adaptation in smaller developing countries

Relative to developed countries, populations in developing nations are more vulnerable to climate change and will be more adversely affected by its impacts. Their development conditions and economic resource constraints often exacerbate their economic and climate change vulnerabilities, and inhibit their ability to adapt to climate change in social, technological and financial terms. The impacts of climate change will have far-reaching effects on the development of such countries, including their attainment of the Millennium Development Goals and other internationally agreed development objectives by 2015, and in decreasing the ever-widening development gap.

The global projected impacts of climate change are severe and vary by region. Africa, where the majority of least developed countries (LDCs) are located, is projected to be hard hit by increased water-related stresses such as droughts in large parts of the continent, which will severely compromise food production and security. Projected sea-level rise is likely to affect low-lying coastal areas with large populations. Many parts of developing Asia will likely see decreased freshwater availability, and many low-lying coastal areas with large populations are likely to face increased flooding from sea surges or flooding rivers.

In Latin America, projections are that the Amazon River Basin will start drying out by mid-century, turning from tropical forest to savannah. Agricultural productivity is projected to decrease, and water availability could also be significantly affected. Small island developing states (SIDS) are expected to be most adversely affected by sea-level rise exacerbating inundation, storm surges, erosion and other coastal hazards, threatening vital infrastructure and the livelihoods of island communities. The availability of coastal resources (such as fisheries) is projected to be adversely affected due to, inter alia, beach erosion and coral bleaching. Water availability in many small islands in the Caribbean and the Pacific is also expected to become insufficient to meet demand during low-rainfall periods by mid-century.

LDCs: trade profiles and potential climate change effects

Many LDCs are being left behind other countries in terms of trade growth and competitiveness, while the quality of their trade growth is neither sustainable nor equitable. LDCs' export profiles are typically characterised by high degrees of export concentration, with only a few product tariff lines, such as mineral fuels, oil and clothing being exported. These profiles render them highly vulnerable to market demand-side volatility and other shocks – such as climate shocks – in relation to their main export products.

Climate change impacts on LDCs' trade capacity and competitiveness are likely to vary depending on the product and the country concerned. While LDC oil exporters may benefit from high fuel prices, possible oil price adjustments by oil importing countries as a result of climate change-related policies may lower demand for fossil fuels and thereby reduce LDC oil exporters' future export earnings.

Other commodity exports of LDCs that are highly dependent on climate conditions, such as fisheries and agricultural products (e.g. cotton, coffee, tea, mate and spices, tobacco, vegetables, seed oils, fruits, animal skins) could also be adversely affected due to, for example, the reduction of growing seasons, drought and ecosystem loss.

Furthermore, the development and maintenance of trade-related infrastructure, such as roads, railways and ports, might also become more difficult as countries struggle to match increasingly scarce resources with increasing climate adaptation demands.

SVEs: Trade profiles and potential climate change effects

Their limited physical size and constrained natural resource and labour endowments often mean that the domestic markets of small vulnerable economies (SVEs) cannot support the location of large-scale industries or the production of goods subject to economies of scale at competitive prices. This means that the range of products produced in SVEs is often limited or products are not priced competitively. SVEs hence often show a very high dependence on imports and exports and, consequently, on foreign market conditions with trade to GDP ratios usually much larger than the average for other developing countries and with exports generally relying on a very narrow range of goods and services, concentrated on the markets of a few countries. In many SVEs, these factors create a high degree of trade dependence and economic instability, which both affect and are affected by their vulnerability to climate change.

The services sector, especially tourism, has become a major source for economic activity in many SVEs. Some SVEs, especially among the small island developing states (SIDS), have sought to overcome economic instability and their associated trade-related challenges by pursuing economic diversification strategies to become knowledge and service-based economies. For other SVEs, however, agriculture remains the dominant economic activity to supply both domestic and export markets.

Adapting to climate change: the role of development-oriented trade policy

Adapting to climate change entails taking the right measures to reduce the negative effects of climate change (or to exploit the positive ones) by making appropriate adjustments and changes. There are many options and opportunities to adapt, ranging from technological options to behavioural changes at the individual level, all of which would depend on the policy choices taken by individual countries, whether unilaterally or in the context of their international commitments and obligations.

The concept of adaptation to climate change is closely linked to the development of adaptive capacity, which refers to changes in processes, practices or structures to moderate or offset potential damages or to take advantage of opportunities associated with changes in climate. National-level adaptation by developing countries will require a strong component of international co-operation from developed countries for support in meeting the costs of adaptation, as provided for in the UNFCCC. Finally, effective climate adaptation by developing countries cannot take place unless other climate-adaptive changes are also effected in other areas of international policy, such as trade and intellectual property, to make these fully supportive of developing countries' efforts to undertake sustainable development. In short, effective climate adaptation is premised on the achievement by developing countries of their first and overriding priorities of economic and social development and poverty eradication.

With respect to ensuring the appropriate adaptation of trade policy to climate change impacts and constraints that may arise in a developing country context, especially for smaller developing countries, it will be necessary for policy-makers to rethink their current macroeconomic development and trade policy. The trade challenges that these vulnerable countries face stem in large part from the confluence of economic and trade liberalisation policy reforms undertaken in the past three decades, which have to a great extent exacerbated these countries' vulnerabilities to both economic volatility and climate change.

Rethinking trade policy approaches in light of climate change

In light of their trade-related challenges and the impacts that climate change is having and will have on the trade competitiveness and economic development prospects of smaller developing countries, it is clear that the key for effective climate change adaptation by these countries lies in building domestic economies that are resilient, diversified and more productive, especially in sectors that are not as vulnerable to climate change impacts.

Further trade liberalisation by LDCs and SVEs in the sense of further increasing levels of tariff openness to imports may not necessarily be the appropriate 'climate-adapted' trade policy response. Rather, adjustments are needed to the trade policies and measures of their trading partners in order to adequately reflect and respond to the trade, development and climate change challenges that smaller developing countries face.

In short, what would broadly be required by smaller developing countries with respect to ensuring that their trade and development policies and measures are climate adapted are: (i) domestic sustainable development policies that incorporate trade and other economic measures designed to build up domestic productive capacity and promote economic diversification to sectors and activities that are less vulnerable to climate change than current ones; and (ii) an enabling and coherent international policy environment (in both the trade and climate regimes), which includes economic and climate policy space and flexibility, new and additional financial flows to support climate adaptation and development, research and transfer of climate-friendly technologies, and external support aimed at improving these countries' trade competitiveness and long-term climate-adapted sustainable development.

With respect to trade in goods, climate-adapted trade-related policy-making that is oriented towards sustainable development could imply a rethinking of LDCs' and SVEs' agricultural development policy and programmes in order to take into account climate change-related impacts. This could involve promoting shifts in both crops grown and production processes used to factor in climate change-related stressors. Domestic policy shifts may need to be effected – e.g. prioritising agricultural production for food security and industrial production of manufactured goods for domestic consumption over export-oriented production and as a means to lessen vulnerabilities to external trade, economic and environmental shocks.

Attention should be paid to the ancillary agricultural and industrial policy and infrastructure shifts that may need to be made to support diversification away from the current agricultural commodities being primarily exported by smaller developing countries. Adapting to climate change impacts on infrastructure and settlements could include scientific services to assess vulnerabilities, retrofitting buildings, raising awareness and establishing resettlement programmes.

Climate adaptation will also require that trade-related infrastructure (such as ports, roads, etc.) be built or strengthened in preparation for climate impacts. Adaptation should include, where appropriate, the development of alternative infrastructure if existing ones cannot be climate adapted.

As with LDCs and SVEs primarily dependent on the agricultural commodities export sector, those dependent on their oil and other mineral commodity exports will need to ensure that income gains coming from high prices for oil commodities are invested into improving the diversification level of their economies. This could be achieved by their investing in other productive economic sectors and providing sufficient resources to implement a strategic industrial development policy.

For many SIDS (both LDCs and SVEs), trade in products from their coastal zones and fisheries are major components of their economic and trade profiles. Climate change impacts on these fragile ecosystems, while difficult to combat, could be better addressed through more robust domestic regimes of resource access, control and management.

Climate-adapted agricultural and industrial diversification will also require securing adequate sources of energy to fuel existing production and expansion into new economic activities while, at the same time, not contributing more than is necessary to global greenhouse gas emissions. The development of domestically sourced, clean, renewable, sustainable energy sources and infrastructure is therefore an important component of climate-adapted agricultural and industrial diversification.

Economic diversification into the services sectors for smaller developing countries may provide many climate-adaptation and development benefits. Services sector expansion and diversification in a manner that is climate adapted will, however, require not only increased domestic public and private sector investment into upgrading the country's human resources, but also investments in improving and climate-proofing the associated infrastructure for specific services sub-sectors and their various modes of supply.

The role of trading partners

The major trading partners of smaller developing countries should support climate-adapted agricultural and industrial sector diversification in these countries, including but not limited to the provision of financial and technology transfer assistance for such diversification in the context of adaptation consistent with developed countries' obligations to do so under the UNFCCC and immediate action to remove barriers (both tariff and non-tariff) to these countries' exports.

Developed country subsidies that support unfair competition with the products or production capacity of LDCs and SVEs, whether in agriculture, industrial goods or even fisheries, should be reduced or eliminated.

Support should also be provided in terms of recognising and allowing for smaller developing countries' needs for flexibility, with respect to policy shifts and measures within agricultural and industrial diversification programmes. Any market access opening that smaller developing countries may wish to undertake should be voluntary and flexible to take into account possible changing economic and climatic conditions.

The trade-related actions above should cohere with and enhance the mandatory actions that need to be taken by developed countries under the UNFCCC, with respect to the provision of financing and technology transfers to developing countries. Such actions will determine the extent to which developing countries will be able to effectively implement their own UNFCCC commitments.

Finally, in relation to the World Trade Organization (WTO) negotiations, a genuine effort to contribute on trade and climate issues would involve responding effectively to developing country proposals for an integrated and development-oriented approach in determining the negotiated outcomes of the Doha Development Agenda.

Specific Trade and Climate Change Concerns

Bunker and aviation fuels

As the transport sector comes under pressure for action on climate change mitigation, there will be moves towards more integrated transport systems with logistical approaches, traffic management and urban planning becoming increasingly significant. Internationally, policy approaches to emission reductions in the aviation and shipping sectors are likely to be based around carbon taxes and environmental levies, with potentially significant consequences on the long-haul tourist destinations and the cost of international trade which LDCs, SVEs and SIDS are particularly dependent on.

The concerns of smaller developing countries in this area are complex. For trade-dependent and remote or landlocked LDCs, SVEs and SIDS, transport costs are an important competitiveness concern. Many are also highly dependent on the tourism and travel industry, which has been an important diversification strategy. The tourism sector is considered to be a major emitter, but it is also highly vulnerable to the effects of climate change through damage to ecosystems, sea-level rise and changes in the patterns of vector-borne diseases.

At the international level, developing countries, and particularly SIDS, recognise their vulnerability to climate change and urge both a focus on adaptation and support from those parties responsible for climate change, which need to take a lead on mitigation. At the same time, SIDS are advancing a proactive agenda that looks at adaptation and mitigation in tandem. This urges the development and application of efficient energy technologies that help mitigate the effects of climate change, and so reduce the need for adaptation, and deliver a financial dividend through a reduced use of imported fossil fuels which can also be diverted towards development and adaptation. SIDS also co-operate at the regional level to respond to the climate change challenge, and work with the international aviation and cruise line industries, as well as NGOs that seek to promote sustainable tourism, to improve their climate profiles.

Current discussions on international levies on bunker and aviation fuels, or an international trading scheme in this area, need to consider the low emission levels of LDCs', SVEs' and SIDS' economies, as well as the impact that levies might have on these and apply the principle of common but differentiated responsibilities in any response.

Carbon labelling and the food miles debate

The world trade in fresh fruits, fruit products, vegetables and vegetable products has more than doubled over the past two decades between 1982–1984 and 2002–2004.

Developing countries export a third of the total trade, and the social and economic benefits of this trade are crucial in terms of income and jobs.

Consumer concerns over the environmental impact of airfreighting resulted in the first forms of carbon labelling in supermarkets, with aeroplane stickers used to indicate fresh produce that had been airfreighted. The stickers, however, single out just one part of the carbon footprint of airfreighted goods: their transport from country of origin to country of sale. They ignore other parts of the process, including the footprint of the production process and road-transport factors, giving a distorted picture.

The role of voluntary carbon-labelling schemes is likely to grow in the future, providing consumers with the option of decreasing their personal carbon footprints. Given this context, the debate on food miles needs to be expanded to not just include road and sea transport but to look at the total carbon emissions of a product through the supply chain, using life-cycle analysis, and evaluate how to reduce emissions at each stage of the chain to achieve a carbon neutral rating.

Overall, carbon-labelling schemes provide opportunities as well as challenges for developing countries. Any future carbon schemes need to balance the provision of useful and accurate data with the need to be simple and transparent and ensure low transaction costs for small country producers. Many producers are concerned that labelling and standards are becoming barriers to market access. They see the rise in such 'non-tariff barriers' as potential obstacles to market entry and as a vehicle for green protectionism.

Many developing countries also feel that their voice is not heard in the development of private sector-labelling schemes and have limited room to manoeuvre since WTO rules in this area bind member countries, not private organisations. Small developing countries need to engage strongly with the private sector and non-governmental organisations on standard setting and labelling issues.

Agriculture, fisheries and forestry

Agriculture contributes significantly to the economy of many developing countries. On average, agriculture contributed about 2 per cent of GDP in developed countries in 2004, but 11 per cent in developing countries, and an average of 40 per cent in Africa. Global agricultural trade, and rules governing this trade, affect carbon management globally, as changes in land use patterns have major impacts on the carbon balance. Agriculture and food security are also highly vulnerable to climate change. This vulnerability is not only caused by the vagaries of climate, but also by the climate's interaction with other developmental challenges. The impacts of climate change on trade will involve changes in comparative advantage based on environmental factors.

Both mitigation and adaptation measures can and should be taken within the agriculture sector. On the mitigation side, these include: cropland management, grazing land management and pasture improvement, restoration of degraded lands, livestock management, manure management and bioenergy. Many of these approaches provide win-win outcomes or co-benefits in terms of higher productivity, better management of natural resources and habitats, or the production of valuable by-products. Others require substantial investment at the global level, such as the development of low-emission rice varieties and livestock breeds. The options are not universally applicable and countries should evaluate and apply those best suited. In addition, possible negative effects or trade-offs of the measures should be considered.

Adaptation strategies include farm-level practices such as planting different varieties, changing planting dates, crop and livestock diversification, adapting to a shorter growing season, rotating or shifting production between crops and livestock, and shifting production away from marginal lands. Macro-level adaptation strategies include the introduction of insurance and other financial mechanisms to mitigate risks, as well as temporary migration where migrant farmers relocate from drought-affected areas to more favourable regions to farm, and subsequently return to their villages when conditions improve.

Adaptation to longer-term climate change is still in its infancy, despite 'showcase' examples of strategies. These strategies are often implemented as a planned adaptation and include the development and adoption of new technologies and strategies to reduce greenhouse gas emissions, as well as build the resilience of the agricultural sector. It is noteworthy that poorly designed and implemented adaptation strategies can result in maladaptation. This is where adaptive responses result in unintended, adverse, secondary consequences that outweigh the benefits of undertaking the strategy. There are also limits to adaptation, the most visible of these being finance, but others include the loss of some ecosystem species, human systems and cultures.

The global community must support adaptation in the agriculture sector in developing countries. This also needs to be reflected in global trade negotiations. Trade justice should be promoted by ensuring that governments, particularly in poor countries, can choose the best solutions to end poverty and protect the environment. These may not always be free trade policies.

Climate change and trade policies will also impact significantly other natural resources sectors such as fisheries and forestry. In these sectors, as with tourism, sustainable approaches to development, production and harvesting are key determinants of resilience and need particular attention. There is also an increasing concern on the part of consumers about the sustainability of fish and timber harvesting. Addressing these concerns could be an important strategy for smaller developing countries in securing and maintaining their market share in these sectors.

Energy efficiency

LDCs, SVEs and SIDS lack access to technology in addressing their climate change concerns, and they face significant availability, cost and trade barriers in accessing the technologies they need to improve energy efficiency and develop the use of renewables. These are significant concerns with respect to competitiveness for remote economies, both in terms of cost and a desire to demonstrate clean production methods. Constraints to addressing these challenges include an inadequate awareness in-country of energy-related challenges, a lack of technical capacity for energy audits and needs assessments, and financial resources, since the financial cost of undertaking assessments is often a barrier to small developing countries in addressing their policy concerns.

Energy efficiency improvements provide a particularly cost-effective way to reduce energy bills and greenhouse gas emissions. However, energy-efficient end-use technologies are relatively underutilised because of market barriers, which can be overcome in part by direct policy measures such as minimum energy performance standards and energy labelling. As with carbon labelling, any international energy standards and labelling schemes need to balance the need for accurate and useful data with the need to be simple, transparent and involve sufficiently low transaction costs to include small countries and players.

Environmental goods and services

The rapid diffusion of clean technologies will be a key to climate change mitigation. Developed country producers and exporters have proposed, at the WTO, to single out environmentally friendly technologies, and among these, climate-friendly technologies, and ask countries to bring down tariff barriers to aid their diffusion. However, developing countries have been reluctant to make such blanket commitments. For smaller developing countries, the priorities in this area are mixed. LDCs, SVEs and SIDS often have high tariffs on environmental and energy-efficient goods and tariff cuts could have significant implications for government revenues. Furthermore, technologies that relate to adaptation – which is a key priority for LDCs, SVEs and SIDS – have not been a priority in discussions on the liberalisation of environmental goods and services, because their end-use is difficult to define and they are often low-tech and locally provided. Overall, trade liberalisation by itself may not be sufficient to promote the diffusion of climate-friendly goods and services, and a host of complementary measures may be needed, including regulatory, capacity-building, financial and technology related measures.

The development and transfer of technology for climate change mitigation and adaptation has also emerged as a basic building block of the current climate change negotiations. A wide range of technology needs have been identified, together with barriers in access to those technologies, but there will likely be no single answer to promoting the transfer of climate-related technologies.

The UNFCCC has recognised that least developed countries, countries in the African region and small island developing states have 'specific needs and special situations' with regards to technology transfer. Many of the tools found useful in other contexts may be inadequate and even counterproductive for these countries, which have lower levels of development and other distinctive circumstances. For example, market-based technology transfer mechanisms such as foreign direct investment or joint ventures are ineffective in meeting the needs and demands of least developed countries. There is increasing realisation that – both within and beyond the intellectual property system – existing innovation structures and activities can and should be enhanced, and more efforts are needed to get smaller developing countries on board.

Border measures and energy-intensive sectors

Climate-related border measures to address free-rider/competitiveness concerns with respect to goods produced in countries with climate mitigation measures in place, and those without, have been under consideration by the EU and US and are highly controversial. These measures have three aspects: the form that the measures would take; the industries covered (energy-intensive sectors); and the source countries to be covered.

There has also been renewed interest in sectoral agreements covering energy-intensive and heavily trade-exposed industries such as steel, pulp, paper, aluminium, cement and agrochemicals. Proponents argue that a sectoral approach should be considered as part of a larger international framework on climate change mitigation and could alleviate competitiveness concerns while facilitating the diffusion of best-available technologies and engaging major emitters through technology transfer and financing.

While small developing countries do not face mandatory emissions cuts, given their high level of natural minerals exports, it could be strategic in the long run for developing countries to adopt sectoral emissions targets that would help attract technology and finance support across a bundle of small-scale projects in an individual sector. Given that many are resource-rich, small developing countries could view emissions mitigation under sectoral approaches as an opportunity to upgrade export capacity from raw mineral products to finished goods. Participation in a sectoral approach regime could mark the entry of resource-rich small developing countries into energy-intensive industries. The use of 'best-available technologies' through the sectoral initiative conforms domestic production to international standards, essentially linking an individual market to the global value chain. Participation could be helpful to small developing countries if they use backward linkages and capacity building in keeping with broader, sustainable development goals.

1

Introduction: Key Issues for Smaller Developing Countries

Ivan Mbirimi, Independent Consultant

This chapter is intended primarily as an introductory guide to the debate on trade, climate change and sustainable development – a bird's eye view of the linkages between climate change, on the one hand, and trade and sustainable development on the other, with some guideposts as to what might constitute an effective response. It provides in a non-technical and over-arching manner, a summary of the likely impacts of climate change on development in least developed countries (LDCs), small and vulnerable economies (SVEs) and small island developing states (SIDS) – that is, smaller developing countries (SDCs) – and how some of the impacts might affect the trade of these countries. This discussion is of vital importance to these smaller developing countries because as Paul Roberts has noted, 'Worse, climate change is not an equal-opportunity disaster. Whereas the northern, and richer countries might suffer relatively minor detriment or might even benefit in certain ways from global warming, the severest effects will be felt disproportionately in Africa, in parts of Asia and among some of the tiny island states'.

Climate change and development

LDCs, SVEs and SIDS (see table 1.1) have contributed little to climate change, yet they are among the countries expected to suffer its severest effects – extreme rainfall variability and crop failures; frequent severe floods, droughts and storms; and rising sea levels. For many of these countries, climate change is already a real and present danger, the consequences of which they are struggling to contain. However, these impacts capture only one side of the story – the risk or exposure element. Far more important in development terms is the vulnerability element, which refers to a country's capacity to manage external shocks without undermining long-term development.

The distinction between risk and vulnerability is an important one: for while communities are all exposed to the risk of climate change (albeit to varying degrees), their vulnerability varies according to degree of exposure of each community and its coping capacity. This exposure is likely to be greater in a community at a low level of

Table 1.1 LDCs and SVEs, including SIDS

Least developed countries[1]	*Small vulnerable economies*[2], *including small island developing states*[3]
1. Afghanistan	1. Albania
2. Angola	2. Antigua and Barbuda
3. Bangladesh	3. Armenia
4. Benin	4. The Bahamas
5. Bhutan	5. Bahrain
6. Burkina Faso	6. Barbados
7. Burundi	7. Belize
8. Cambodia	8. Bolivia
9. Central African Republic	9. Botswana
10. Chad	10. Brunei Darussalam
11. *Comoros*	11. Cameroon
12. Dem. Rep. of Congo	12. Cape Verde
13. Djibouti	13. Cuba
14. Equatorial Guinea	14. Dominica
15. Eritrea	15. Dominican Rep.
16. Ethiopia	16. Ecuador
17. The Gambia	17. Fiji
18. Guinea	18. FYR Macedonia
19. *Guinea-Bissau*	19. Gabon
20. *Haiti*	20. Georgia
21. *Kiribati*	21. Ghana
22. Lao PDR	22. Grenada
23. Lesotho	23. Guatemala
24. Liberia	24. Guyana
25. Madagascar	25. Honduras
26. Malawi	26. Jamaica
27. *Maldives*	27. Jordan
28. Mali	28. Kenya
29. Mauritania	29. Kyrgyzstan
30. Mozambique	30. Macao China
31. Myanmar	31. Marshall Islands
32. Nepal	32. Micronesia (Fed. States of)
33. Niger	33. Mauritius
34. Rwanda	34. Moldova
35. *Samoa*	35. Mongolia

36. *São Tomé and Príncipe*
37. Senegal
38. Sierra Leone
39. *Solomon Islands*
40. Somalia
41. Sudan
42. *Timor-Leste*
43. Togo
44. *Tuvalu*
45. Uganda
46. United Republic of Tanzania
47. *Vanuatu*
48. Yemen
49. Zambia

36. Namibia
37. Nauru
38. Nicaragua
39. Palau
40. Panama
41. Papua New Guinea
42. Paraguay
43. St Kitts and Nevis
44. St Lucia
45. St Vincent and the Grenadines
46. Singapore
47. Sri Lanka
48. Suriname
49. Swaziland
50. Tonga
51. Trinidad and Tobago
52. Uruguay
53. Zimbabwe

---

development, while coping capacity is enhanced the greater the capacity to manage risk. In practice, how climate change risks combine with the vulnerability of a particular location will determine the magnitude of climate change impacts for that location. For smaller developing countries, their high vulnerability is one of the main reasons why they are expected to shoulder a disproportionate share of the costs of climate change.

Least developed countries are vulnerable because they are poor. Their poverty – in terms of both income and resources – translates into a limited capacity to adapt to external shocks and hazards of all sorts, including climate change, as well as to benefit from opportunities that may arise, for example from modern, friendly technologies. Their poverty is also the main reason for their continued marginalisation in world trade. The most recent UN Conference on Trade and Development (UNCTAD) Report on LDCs (2008) points out that despite an above-average export performance, LDCs continue to be marginalised, mainly because of their large populations (relative to resources) and other factors which mean that even in relatively good years, such as 2005–06, their per capita income still lags behind that of other developing countries. Furthermore, in relation to their participation in international trade, their performance has been hampered by a variety of supply bottlenecks – inadequate production, poor infrastructure, poor quality products and low technology. Climate change is likely to exacerbate these supply and competitiveness constraints.

Viewed in this context, climate change is a massive additional constraint on development and one with a huge potential to undermine economic growth and development, because it is likely to affect economies across the board. Superimposing climate change impacts on existing development constraints is likely to further complicate the development task for these countries. It will also make diversification out of agriculture – upon which many of these countries depend – more necessary, because many crops grown in LDCs, particularly African LDCs, are being produced at or near the limit of their thermal tolerance. Climate change must therefore be approached primarily as a development challenge. It will present developing countries with many of the same challenges and demands that they currently face, but on a larger scale. In fact, coping with climate change impacts may require a kind of 'structural adjustment' involving resource (labour and capital) shifts away from agriculture and into other more viable sectors and activities. The problem of course is that these resource shifts will not take place seamlessly. They will require massive investments in infrastructure and strengthening of human capacity, including the management of risk and the creation of a regulatory framework that emphasises flexibility and adaptability to encourage the necessary shift of resources into viable sectors of the economy. It would seem that under these circumstances, the most important task for governments might be to provide appropriate incentives, adequate information and an economic environment conducive to investment in the required changes. This, in a nutshell, is what the debate about adaptation is about.

The adaptation debate, however, must be informed by past experience. The simple fact is that LDCs, particularly in Africa, have the poorest record of structural adjustment. Over the last half-century, they have not displayed a high degree of adaptability, with the result that their economies remain structurally the same in terms of the contribution of different economic sectors to GDP, as well as their main export products. Furthermore, should these countries be forced out of agriculture by climate change, the alternatives include industrialisation, a sector they have previously failed to diversify into and in which they would now face an additional hurdle in the form of intense competition from East Asian countries, including China.

Another lesson from experience – one even more relevant to climate change impacts – is that developing countries are not a uniform set of countries that will all experience climate change in exactly the same way. Even Africa, which is one land mass, will experience climate change differently. Uniform policy responses are therefore unlikely to work. This has one immediate implication: individual developing countries need to quickly build an understanding of the likely impacts of climate change on their economies, which means collecting the necessary information and data. While it will take some time for many of these countries to assemble reliable information on climate and weather patterns, a start must be made immediately.

What this experience suggests is that some of the key impediments to adaptation to climate change have dogged developing countries for a long time. They include small and fragmented markets, a poor policy and business environment, weak or non-existent infrastructure, poor uptake of technology and a high dependence on one or two export commodities. These constraints must be addressed alongside those brought about by climate change. Some of the solutions include more effective regional co-operation, a policy that developing countries have tended to pay only lip service to.

Climate change and trade

One way of encapsulating the links between trade and climate change for smaller developing countries is to distinguish between direct and indirect effects (see figure 1.1). Direct effects include climate change impacts on production of goods and services – for example, when climate change necessitates a shift from production of one crop to another that may be more drought-resistant or a shift out of agriculture as a whole. This kind of impact would generally mean reduced supply of an agricultural product previously produced in the country. It would also mean a loss in comparative advantage and, given what we know about LDCs, it would mean a worsening of the supply constraint problem that has been the bane of these countries for many years.

Indirect effects generally stem from actions taken by developed countries as part of their mitigation responses to climate change. They include a range of policies covering carbon labelling, energy-efficiency standards, the so-called 'cap-and-trade'[4] systems and aspects of the clean development mechanism (CDM). These mitigation responses will impact on Africa in both positive and negative ways. The negative impacts, especially those associated with carbon labelling and standards, arise in part because developed countries may be tempted to apply them in a protectionist manner. In addition, they may also have negative consequences because, in the absence of universally accepted

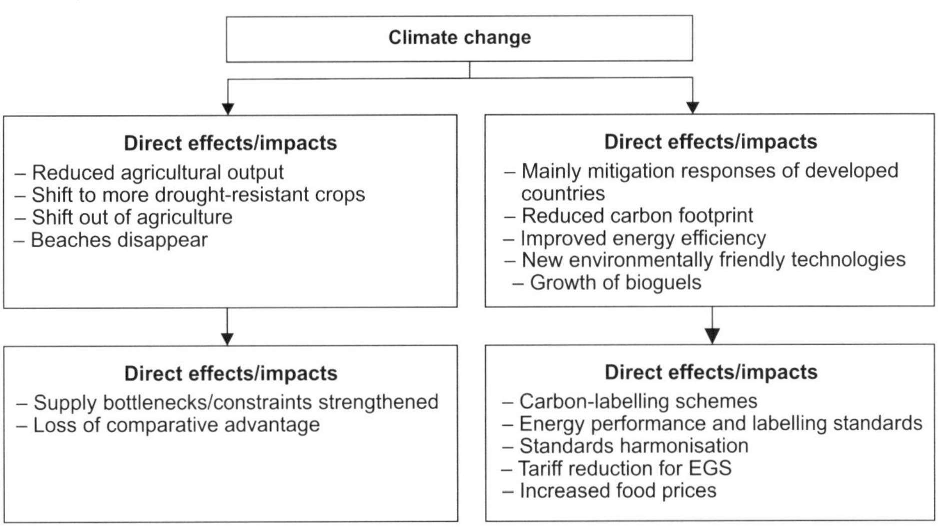

Figure 1.1 Climate change impacts on trade of developing countries

methods of measuring embedded carbon in products and the unavoidable costs of implementing them, their adoption by developing countries may impose costs on small farmers from low-income countries. The positive impacts include opportunities created by new technologies, as well as the fact in circumstances where resources are provided, harmonising standards upward may in fact improve the competitiveness of developing countries.

Without going into the detail of how developing countries might be affected in each of these cases, it is safe to say the balance of costs and benefits will likely vary from country to country. However, all developing countries need to be fully engaged in international negotiations or discussion of these issues, so that they can advance and defend their interests.

Important issues for smaller developing countries

Climate change is a present and real danger for LDCs, SVEs and SIDS, and all the available evidence points to the fact that the burden of climate change will fall disproportionately on them. The fear of negative consequences is not just based on the fact that these countries are located in parts of the world expected to experience major changes in climate, rather it derives from the realisation that adaptation will be inadequate, so that projected risks are likely to be fully realised while any potential opportunities may be missed. It also reflects the concern that superimposing climate change problems on the existing rather intractable development challenges faced by these countries will further complicate the development process.

The most immediate task for governments is therefore to develop adaptation plans that seek to build the capacity and resilience of national economies, as well as mechanisms for identifying potential opportunities. To be effective, these adaptation plans must be backed by relevant evidence. Yet evidence of the likely impact of climate change on specific countries and sectors of the economy is not presently available. It is suggested that countries begin work on this most urgent of tasks: gathering local evidence and building databases of information on climate change immediately. Quite clearly, climate change has the potential to undermine economic growth and the livelihoods of tens of millions of people who live in the fragile ecosystems of LDCs, SVEs and SIDS. This leads to an obvious implication: climate change must be placed at the centre of all development plans and processes.

In terms of climate change impacts on trade, two broad consequences have been identified: those associated with the direct loss of comparative advantage and likely reinforcement of supply-side constraints caused by climate-induced shifts in production structures; and the indirect impacts which are in the main by-products of mitigation responses by developed countries. Many of the trade rules in this area are evolving, which makes it vital for smaller developing countries to be fully engaged in the ongoing debates about setting standards, rules and regulations. In almost all cases, developing countries face two risks. One concerns measurement issues or the yardstick on which a particular standard is to be based. It is important that conditions in developing countries are given due consideration in any discussion of international rules and standards. The other risk relates to the costs of implementing new rules and standards, especially if these are based on developed countries' conditions.

Notes

1. Based on the UN list of LDCs after the 2006 triennial review. See: http://www.unohrlls.org/en/ldc/related/62/ [last accessed 3 March 2009]. LDCs which are also SIDS are indicated in italics.
2. For the purposes of this book, the criteria for determining SVEs specified in paragraph 141 of the Revised Draft Modalities for Agriculture, TN/AG/W/4/Rev.2, 19 May 2008, is applied. The list of SVEs falling within such criteria is contained in Annex I thereof. Footnote 10 of the draft modalities text indicates that Congo, Côte d'Ivoire and Nigeria may also be eligible for SVE treatment as well as other WTO Members 'that can provide data that show that they can meet the criteria in' paragraph 141.
3. The list of SIDS is based on the UN High Representative for Least Developed Countries, Land-Locked Developing Countries, and Small Island Developing States, at http://www.unohrlls.org/en/sids/44/ [last accessed 3 March 2009]. SIDS are indicated in italics.
4. 'Cap-and-trade' is an approach to controlling GHG emissions by setting through regulation an overall limit (cap) for a specific time period, and then allocating permits to emit a certain quantity of emissions to individual parties (industrial sectors or companies). Those with low emissions may sell unused permits, while those with higher emissions may buy them to help meet their quota.

2

Responses to Trade and Adaptation Challenges

Vicente Yu, South Centre

Achieving sustainable levels of development, characterised by conditions of economic and social equity, remains the fundamental foundation for undertaking effective societal responses to trade and climate change adaptation.

Article 3.4 of the UN Framework Convention on Climate Change (UNFCCC) recognises Parties' right to promote sustainable development, and further goes on to stress that 'policies and measures to protect the climate system against human-induced change should be appropriate for the specific conditions of each Party and should be integrated with national development programmes, taking into account that economic development is essential for adopting measures to address climate change'. The balance of commitments among Parties to the UNFCCC reflected in Article 4.7 further states that implementation of UNFCCC commitments 'take fully into account that economic and social development and poverty eradication are the first and overriding priorities of the developing country parties'.

In the same vein, the World Trade Organization (WTO) Agreement in its preamble also explicitly indicates that sustainable development is an institutional objective. This preambular statement, according to the WTO Appellate Body in the US–Shrimp Turtle case, is supposed to give 'colour, context and shading to the rights and obligations of Members under the WTO Agreement, generally, and under the GATT 1994, in particular'[1].

More generally, Principles 3 and 4 of the Rio Declaration on Environment and Development refer to the right to sustainable development.

The relationship between trade and climate change measures in the climate regime is governed by, among others, Article 3.5 of the UNFCCC, which states that 'measures taken to combat climate change, including unilateral ones, should not constitute a means of arbitrary or unjustifiable discrimination or a disguised restriction on international trade'. This language, in fact, is taken from the chapeau of Article XX of the General Agreement on Tariffs and Trade (GATT), which allows WTO Members to adopt measures that may be inconsistent with their WTO obligations if such measures are, inter alia, 'necessary to protect human, animal or plant life or health', provided that these measures 'are not applied in a manner which would constitute a means of arbitrary or unjustifiable discrimination between countries where the same conditions prevail, or a disguised restriction on international trade'.

Policy approaches to trade and environment challenges – including on climate change adaptation – are therefore premised under both the UNFCCC and WTO regimes on a clear recognition of the right to development and the need to ensure that the right is promoted and effectively achieved. Maintaining the focus on promoting and achieving the right to development, especially development that is sustainable, is therefore essential for meeting the objectives of both the climate regime under the UNFCCC and the trade regime under the WTO. In doing so, trade measures (including unilateral ones) that may be imposed to combat climate change must not be arbitrary or unjustifiable discrimination or a disguised restriction on the international trade of developing countries, especially least developed countries (LDCs) and small vulnerable economies (SVEs), including small island developing states (SIDS)[2].

The development challenge

Major development challenges continue to exist, and addressing these continues to be the overriding priority of developing countries. While developing countries as a group (including China and India) achieved an average of 5–6 per cent growth between 2002 and 2007, '... not all countries or segments of the population are beneficiaries of this growth...'[3]. In addition, some large developing countries such as China and India are now 'engines of growth for the world economy ... [and] the share of South–South trade is increasing in the world economy, making inter-South trade a veritable locomotive of growth'[4].

Both the United Nations and the World Bank project continued global economic growth, albeit at a slower rate, over the short-term, largely as a result of the continued expansion of developing country economies[5]. However, systemic global economic inequality is likely to persist in the medium- and long-term, with most developing countries continuing to remain at low levels of economic development, especially in Africa, South Asia and the Pacific.

Income inequality between countries remains extremely high, and whatever income convergence with developed countries might take place will likely be concentrated in only some developing countries rather than being broad-based across all developing countries[6]. Even when developing countries have higher growth rates, the absolute income gap with developed countries on a per capita purchasing power parity basis will continue to increase 'precisely because the initial income gaps are so large ... If average incomes grow by 3 per cent in sub-Saharan Africa and in high-income Europe, for example, the absolute change will be an extra US$51 per person in Africa and an extra US$854 per person in Europe'[7].

The recognition that the development gap was not shrinking led to global initiatives in the early 2000s, which intended to focus global attention on the need to address that gap. However, unless effectively addressed and adapted to, climate change is posing and will continue to pose real-world economic, natural resource and social constraints that could result in a derailment of the accelerating development process of fast-growing developing countries, and perhaps even forestall many other developing countries from embarking on a sustainable development path. Effective adaptation to climate change by developing countries, enabled by co-operation in financing and technology transfer from developed countries, and coupled with a more aggressive greenhouse gas emissions reduction regime for developed countries, is required.

The climate change challenge

Anthropogenic global warming

Global warming is 'unequivocal', as stressed by the Intergovernmental Panel on Climate Change (IPCC) in its Fourth Assessment Report[8]. The temperature increase is globally spread in terms of both land and ocean surface temperatures, with higher levels of increases in high northern latitudes[9]. Sea-level rise is occurring, in large part (57 per cent) due to ocean thermal expansion, followed by melting of glaciers and land-based ice caps (28 per cent) and polar ice sheets[10]. Precipitation changes are occurring, with more precipitation in eastern parts of North and South America, northern Europe and northern and central Asia, and declining precipitation in the Sahel, the Mediterranean, southern Africa and parts of southern Asia (where many developing countries are located)[11]. Extreme weather events (such as heat waves, heavy precipitation events such as storms, and sea surges) have changed in frequency and/or intensity, including observable increases in tropical cyclone activity in many regions[12].

Current global warming is due primarily to anthropogenic emissions of greenhouse gases (GHGs), which 'have grown since pre-industrial times, with an increase of 70 per cent between 1970 and 2004[13]. The largest part of GHG emissions growth between 1970 and 2004 'has come from energy supply, transport and industry, while residential and commercial building, forestry (including deforestation) and agriculture sectors have been growing at a lower rate'[14]. To a great extent, the growth of the global economy and of the global population between 1970 and 2004, both of which drive increased energy-related CO_2 emissions, have resulted in a reversal of 'the long-term trend of declining CO_2 emissions per unit of energy supplied ... after 2000'[15]. On a per capita basis in 2004, developed countries (i.e. those listed in Annex I of the UNFCCC), while having only 20 per cent of global population, 'accounted for 46 per cent of global GHG emissions'[16].

The IPCC projects that 'with current climate change mitigation policies and related sustainable development practices, global GHG emissions will continue to grow over the next few decades'[17] and that 'continued GHG emissions at or above current rates would cause further warming and induce many changes in the global climate system during the twenty-first century that would very likely be larger than those observed during the twentieth century'[18]. Near-terms projections suggest that 'a warming of about 0.2°C per decade' will occur[19].

Projected climate change impacts

The projected impacts of such global warming in the twenty-first century were stated by the IPCC[20] to be as follows:

- Projected warming in the twenty-first century shows scenario-independent geographical patterns similar to those observed over the past several decades. Warming is expected to be greatest over land and at most high northern latitudes, and least over the Southern Ocean (near Antarctica) and northern North Atlantic, continuing recent observed trends.
- Snow cover area is projected to contract. Widespread increases in thaw depth are projected over most permafrost regions. Sea ice is projected to shrink in both the Arctic and Antarctic under all SRES [Special Report on Emissions] scenarios. In some projections, Arctic late-summer sea ice disappears almost entirely by the latter part of the twenty-first century.
- It is very likely that hot extremes, heat waves and heavy precipitation events will become more frequent.
- Based on a range of models, it is likely that future tropical cyclones (typhoons and hurricanes) will become more intense, with larger peak wind speeds and more heavy precipitation associated with ongoing increases of tropical sea-surface temperatures. There is less confidence in projections of a global decrease in numbers of tropical cyclones. The apparent increase in the proportion of very intense storms since 1970 in some regions is much larger than simulated by current models for that period.
- Extra-tropical storm tracks are projected to move poleward, with consequent changes in wind, precipitation and temperature patterns, continuing the broad pattern of observed trends over the last half-century.
- Since the TAR [Third Assessment Report], there is an improving understanding of projected patterns of precipitation. Increases in the amount of precipitation are very likely in high-latitudes, while decreases are likely in most subtropical land regions (by as much as about 20 per cent in the A1B scenario in 2100), continuing observed patterns in recent trends.
- Anthropogenic warming and sea-level rise would continue for centuries due to the time scales associated with climate processes and feedbacks, even if GHG concentrations were to be stabilised.
- If radiative forcing were to be stabilised, keeping all the radiative forcing agents constant at B1 or A1B levels in 2100, model experiments show that a further increase in global average temperature of about 0.5°C would still be expected by

2200. In addition, thermal expansion alone would lead to 0.3 to 0.8m of sea-level rise by 2300 (relative to 1980–1999). Thermal expansion would continue for many centuries, due to the time required to transport heat into the deep ocean.

The best estimates indicate that the planet could warm by 3°C by 2100 and that even if countries reduce their greenhouse gas emissions, global warming will continue. Predictions range from a minimum of 1.8°C to as much as a 4°C rise in global average temperatures by 2100[21].

Higher climate change vulnerability in developing countries

Relative to people living in developed countries, populations in developing countries are more vulnerable to, and are will be more adversely affected by, climate change because they 'have fewer resources to adapt: socially, technologically and financially'[22]. Such impacts will have far reaching effects on the sustainable development of developing countries, including the attainment of the Millennium Development Goals and other internationally agreed development goals by 2015.

In terms of the regional impacts of climate change, the following are some examples of major projected impacts[23]:

- Africa, where most LDCs are located, is projected to be hard hit by increased water-related stresses such as droughts, which could reduce yields from rain-fed agriculture by 50 per cent. This could severely compromise food production and security. Projected sea-level rise is likely to affect low-lying coastal areas with large populations (such as Alexandria, Egypt; Lagos, Nigeria; Abidjan, Côte d'Ivoire[24])[25].
- Likewise, most parts of developing Asia will likely see decreased freshwater availability, and coastal areas with large populations are likely to face increased floodings from sea surges or rivers (such cities as Kolkata and Mumbai, India; Dhaka, Khulna and Chittagong, Bangladesh; Guanzhou, Shanghai, Tianjin and Ningbo, China; Ho Chi Minh City and Hai Phong, Vietnam; Jakarta, Indonesia; Bangkok, Thailand; and Yangon, Myanmar[26])[27].
- In Latin America, projections are that the Amazonia will start drying out by mid-century, turning from tropical forest to savannah. Agricultural productivity is projected to decrease, and water availability could also be significantly affected[28].
- **Small island developing states** (SIDS) are expected to be most adversely affected by sea-level rise exacerbating inundation, storm surge, erosion and other coastal hazards, 'thus threatening vital infrastructure, settlements and facilities that support the livelihood of island communities'. Availability of coastal resources (such as fisheries) is projected to be adversely affected due to, inter alia, beach erosion and coral bleaching. Water availability in many small islands in the Caribbean and the Pacific is expected to become insufficient to meet demand during low-rainfall periods by mid-century[29].

The climate change financing gap

The UNFCCC Secretariat has estimated that by 2030, developing countries will need US$28–67 billion in funds to enable adaptation to climate change, corresponding to 0.2–0.8 per cent of global investment flows, or 0.06–0.21 per cent of projected global GDP, in that year.

However, current global funding for adaptation is a fraction of this figure and access to these funds for developing countries is often lengthy and complex[30]. Additionally, financing flows to developing countries to enable them to undertake response measures to climate change (including technology transfer) also fall far short of what is required (see table 2.1).

Trade and climate change challenges of smaller developing countries

The challenge of ensuring development in the context of climate change is most acute and urgent for LDCs and SVEs, including SIDS. While these countries face climatic and development challenges that other developing countries also face, they have specific circumstances and conditions that make them (LDCs and SVEs) especially vulnerable to having their development prospects adversely hit by climate change.

Trade challenges of LDCs

In the first half of the decade 2001–2010, in which the international community is supposed to focus on supporting LDC growth and development under the 2001 Programme of Action for Least Developed Countries for the Decade 2001–2010 agreed during the Third UN Conference on LDCs in 2001, LDCs were on aggregate falling behind in terms of meeting the GDP growth rate target of at least 7 per cent annually[32]. UNCTAD has estimated that there is 'a strong probability that the incidence of poverty is increasing in these countries'[33]. While seven LDCs[34] are on the road to graduating from LDC status between 2008 and 2013, 36 of the others 'were not meeting any graduating criterion' in 2006, when the LDC list was reviewed by the UN Committee for Development Policy, and that 'prospects for progress towards graduation in the foreseeable future are very slim in nearly seven LDCs out of ten, and remain insignificant in nearly two out of ten'[35].

Trade in goods dominates the export portfolio of the 49 LDCs, with trade in commercial services[36] representing only 12 per cent of the total LDC export receipts on average over the period 2000 to 2006. This is lower than the world average of 19 per cent. LDCs' goods exports have increased since 2003, due largely to the rapid increase in the international prices of oil and mineral products outpacing the growth of other goods (principally agriculture and manufacture) and services exports (figure 2.1)[37].

However, despite the growth of LDCs' export trade from 2000 to 2006, their aggregate share in global world exports continues to be less than 1 per cent (0.9 per cent)[38].

Table 2.1 Comparison of requirements for and availability of financial resources

Funding area	*Current estimates of investments and financial resources needed in developing countries by 2030*	*What is currently available or estimated to be made available to developing countries under the Global Environment Facility (GEF) as an operating entity for the UNFCCC's financial mechanism*
Mitigation	US$176 billion (FCCC/SBI/2007/21, table 5) **US$86 billion a year by 2015** (0.2 per cent of OECD GDP = 1/10 of OECD military expenditures) (UNDP, HDR 2007/08, p.194)	**US$990 million** from the GEF 4th Replenishment for the period 2006–2010, with **co-financing to amount to US$1.6518 billion**[31] (see FCCC/SBI/2007/21, table 1)
Adaptation	US$ 28–67 billion (FCCC/SBI/2007/21, table 3 and para.51)	**US$20.4 million** – GEF Trust Fund: Strategic Priority for Adaptation **US$23.5 million** – Special Climate Change Fund (GEF administered) **US$147.0 million** – Least Developed Countries Fund (GEF administered) **US$80–300 million per year for the period 2008–2012** from the 2 per cent share of the proceeds of annual sales of certified emissions reductions from CDM projects – Adaptation Fund (see FCCC/SBI/2007/21, table 2 and para.62)
Technology transfer Emissions reduction-related technology deployment	US$720 billion (an average of US$24–26 billion per year) – (FCCC/SBI/2007/21, para.93 – no breakdown for developing countries; figures based on International Energy Agency [IEA] estimates)	The GEF estimates that 80–100 per cent of GEF climate change mitigation funding fits the technology transfer definitions used by the Convention (see FCCC/SBI/2007/21, table 2 and para.62)
Deployment of renewables, biofuels and nuclear energy technologies	US$33 billion per year (FCCC/SBI/2007/21, para.94 – no breakdown for developing countries; figures based on Stern Review)	As at April 2007, US$10.7 million were available from the Special Climate Change Fund (SCCF) for the programme for transfer of technology (FCCC/SBI/2007/21, para.90)
Public energy R&D	US$20 billion (FCCC/SBI/2007/21, para.94 – no breakdown for developing countries; figures based on Stern Review)	
National communications under the UNFCCC	The need for resources recognised, but no estimate provided (FCCC/SBI/2007/21, para.109)	US$60.08 million – GEF (FCCC/SBI/2007/21, para.107)
Capacity building	The need for resources recognised, but no estimate provided (FCCC/SBI/2007/21, para.121)	More than US$1.46 billion allocated as at June 2002 – GEF support for capacity-building activities in all its focal areas, but no indication as to future amounts (FCCC/SBI/2007/21, paras.114–116)
Public awareness and outreach	The need for resources recognised, but no estimate provided (FCCC/SBI/2007/21, para.128)	GEF indicated that 'it is not possible to quantify the amount that might have been dedicated to such activities under the GEF' (FCCC/SBI/2007/21, para.126)

Source: South Centre (2008) Financing the Global Climate Change Response: Suggestions for a Climate Change Fund (CCF). SC/GGDP/AN/ENV/3, May 2008, para.6, table 1.

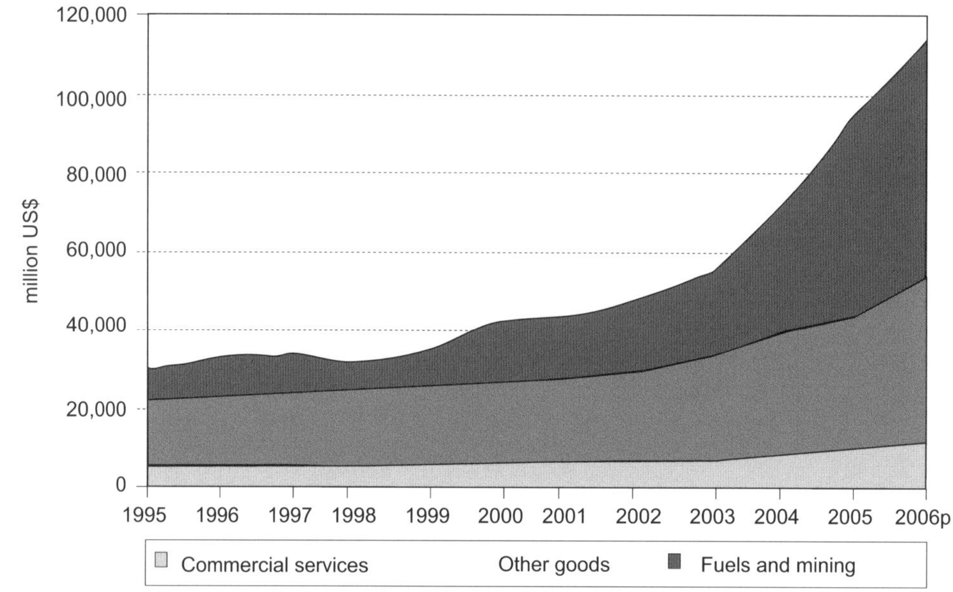

Figure 2.1 Comparative evolution of LDC exports of goods and services, 1995–2006 (million US$)

Note: Estimates based on customs and balance-of-payments data; data for 2006 are preliminary.

Source: WTO, LDC Market Access, para.5, chart 1.

Many LDCs are being left behind in terms of trade growth and competitiveness. At the same time, the quality of that trade growth is neither sustainable nor equitable. LDC oil and mineral fuel commodity exporters' trade has grown more than that for LDCs exporting more agricultural commodities, with the first group also generally tending to now show positive trade balances, while the latter continue to show significant trade deficits[39]. LDC clothing exporters had moderate growth rates, averaging 12 per cent annually over the period 2000 to 2006; however, their export growth has tended to be offset by import growth over the same period, leading to a significant trade deficit[40].

The export profiles of LDCs are typically characterised by high degrees of export concentration, with only a few product tariff lines being exported[41]. Three main products – mineral fuels, oil and clothing – represented almost 50 per cent of total export revenues in 2005 for more than half of the 49 LDCs, and in around 15 LDCs dependence on a few exported commodities is above 90 per cent[42]. As the WTO notes, 'very few LDCs have a diversified export structure that could evenly spread the risks inherent in international market fluctuations'[43]. This is further highlighted by UNCTAD in its latest LDC Report (2008), which stressed that:

> 'International trade is equivalent to over 50 per cent of the GDP of the LDCs as a group, and high rates of export growth have been a key driver of their strong GDP growth performance. However, their export structure remains concentrated on primary commodities and low-skill labour-intensive manufactures. Most LDCs are net food importers. Trade performance is highly dependent on commodity price trends. Trade deficits are increasing in most LDCs, particularly those which specialise in agricultural exports. Despite a high level of integration with the global economy and good export performance, the marginalisation of the LDCs in global trade has declined only slightly if oil is excluded. Marginalisation is rooted in the continuing failure of export upgrading'[44].

LDCs' trade in services, while growing, continues to be a small part of global services trade (1 per cent in 2005, up from 0.4 per cent in 2003)[45]. Their trade in services constituted only 12 per cent of their total trade in goods and services in the period 2000–2006.[46]

The export profiles of LDCs render them highly vulnerable to market demand-side volatility and other shocks – such as climate shocks – in relation to their main export products.

Generally, LDCs are extended non-reciprocal trade preferences for their export goods by their trading partners (particularly developed countries), but 'these preferences are not always utilised by the LDCs'[47]. Despite such preferences, 'excluding arms and oil, the proportion of duty-free exports from LDCs is less than 4 percentage points higher than the average for developing countries (78.7 per cent against 75.3 per cent in 2005)[48]. Furthermore, for the non-mineral fuel and oil commodity exports of LDCs – e.g. agricultural products and textiles and clothing – while LDCs' preference margin over that of other developing countries for agricultural exports is fairly substantial (5.8 per cent), the preference margin with respect to textiles and clothing is 'moderate' (less than 2.5 per cent in both cases)[49]. Finally, the mere fact that privileged market access through preferences exists is often not sufficient to improve LDC export levels, because such levels may depend on factors other than tariffs. For example, non-tariff measures such as sanitary and phytosanitary requirements and technical barriers to trade in importing countries, stringent rules of origin in importing countries, supply-side constraints in LDCs themselves that make it difficult for them to respond to market opportunities, and trade, transport, logistics and administrative difficulties in exporting LDCs may also play a major constraining role in LDC export performance[50].

Climate change impacts on LDC trade capacity and competitiveness

Climate change impacts on least developed countries' trade capacity and competitiveness are likely to vary depending on the product and the country concerned. However, some general observations can be made.

While LDC oil exporters may especially benefit from high fuel prices, possible oil price adjustments by oil importing countries (especially developed countries such as the EU and the US[51]) that may be imposed as a result of climate change-related policies (such as higher carbon taxes on fossil fuels) may serve to drastically lower demand for fossil fuels in their primary markets and thereby reduce these LDC oil exporters' future export earnings.

Furthermore, other commodity exports of LDCs that are highly dependent on climate conditions, such as fisheries and agricultural products (e.g. cotton, coffee, tea, mate and spices, tobacco, vegetables, seed oils, fruits and animal skins), which comprised approximately 9.8 per cent of total LDC exports in 2005, could also be adversely affected. Changing climatic conditions in LDCs (especially in Africa) such as water stress, drought, spreading desertification and ocean temperature variability (for fisheries) could shift temperature and growing conditions in ways that LDCs may find it difficult to adapt to under their current economic circumstances.

Most LDCs are located in Africa – 33 of the 49 LDCs are in Africa, 15 in Asia and the Pacific, and one in the Caribbean – hence the projected impacts of climate change on Africa will be particularly relevant to many LDCs and their populations for purposes of adaptation planning.

The following excerpt from a UNFCCC paper highlights some of the major possible trade-related (in terms of their impacts of trade-related infrastructure and supply-side capacity) climate adaptation challenges that African LDCs are likely to face:

> 'As a result of global warming, the climate in Africa is predicted to become more variable, and extreme weather events are expected to be more frequent and severe, with increasing risk to health and life. This includes increasing risk of drought and flooding in new areas (Few et al., 2004; Christensen et al., 2007) and inundation due to sea-level rise in the continent's coastal areas (Nicholls, 2004; McMichael et al., 2006).
>
> 'Africa will face increasing water scarcity and stress, with a subsequent potential increase of water conflicts as almost all of the 50 river basins in Africa are transboundary (Ashton, 2002; De Wit and Jacek, 2006). Agricultural production relies mainly on rainfall for irrigation and will be severely compromised in many African countries, particularly for subsistence farmers and in sub-Saharan Africa. Under climate change much agricultural land will be lost, with shorter growing seasons and lower yields. National communications report that climate change will cause a general decline in most of the subsistence crops, e.g. sorghum in Sudan, Ethiopia, Eritrea and Zambia; maize in Ghana; Millet in Sudan; and groundnuts in [The] Gambia. Of the total additional people at risk of hunger due to climate change, although already a large proportion, Africa may well account for the majority by the 2080s (Fischer et al., 2002).
>
> 'Future sea-level rise has the potential to cause huge impacts on the African coastlines, including the already degraded coral reefs on the Eastern coast. National communications indicate that the coastal infrastructure in 30 per cent of Africa's coastal countries, including the Gulf of Guinea, Senegal, [The] Gambia, Egypt and

along the East-Southern African coast, is at risk of partial or complete inundation due to accelerated sea-level rise. In Tanzania, a sea-level rise of 50cm would inundate over 2,000km^2 of land, costing around US$51 million. Future sea-level rise also threatens lagoons and mangrove forests of both eastern and western Africa, and is likely to impact urban centres and ports, such as Cape Town, Maputo and Dar Es Salaam'[52].

What the projected impacts of climate change on Africa imply is that the conditions for improving agricultural production in African LDCs for both food security and export are likely to become much more difficult as global warming progresses. Furthermore, the development and maintenance of trade-related infrastructure, such as roads, railways and ports, might also become more difficult as African countries struggle to match increasingly scarce resources with increasing climate adaptation demands.

Table 2.2 Summary of trade and trade-related climate challenges to development and adaptation for LDCs

Key characteristics of current trade profile	*Some projected major climate risks*
• Low global trade share (1 per cent in 2005) • High level of commodity export dependence • High levels of export concentration (mineral fuels and oil; commodity agriculture such as coffee, cotton, bananas, sugar, fish; textiles and clothing); low economic diversification and resilience • Supply-side constraints (infrastructure, trade logistics, finance, information) • Low levels of preference utilisation, due to supply-side constraints, tariff peaks and escalations and demand-side non-tariff measures (NTMs) • Still highly reliant on EU + US markets	• Water stress (drought in Africa, excessive precipitation in Asia) • Sea-level rise (especially affecting low coastal areas in Africa and small island LDCs) • Overall warming (land and water) resulting in changes in crop growth cycles and crops that can be grown; fish stocks in rivers, lakes and seas • Increased human health hazards due to changing disease patterns and vectors
Some potential climate change impacts on trade competitiveness, prospects and development	
• Water stress impacts on agricultural productivity (especially small farmer production) may affect current commodity agricultural export performance • Flooding and desertification may adversely affect production and export activities due to damage to productive and transport infrastructure • Sea-level rise may result in damage or loss of coastal infrastructure, including ports and roads (especially in many island LDCs), hampering trade transport logistics and increasing transport costs • Human health impacts may affect labour force availability and productivity adversely, affecting export performance and economic diversification potential • Climate pressure may spur diversification to other crops, products or services for export that are more appropriate to new climate conditions	

Asian LDCs, of which eight are located on the continent and seven are small island developing states, are also likely to face major climate stresses that may adversely affect their trade supply-side capacity and competitiveness in the global textile fibres, fabrics and clothing trade, including increased rainfall and flooding and sea-level rise [53]. These impacts could imply that productivity increases with respect to the agricultural and manufacturing sectors of mainland Asian LDCs (especially Bangladesh, Nepal and Cambodia) – especially in those sectors of their export interest (such as textiles and apparels) – may be hampered by an increased frequency of water-related extreme weather events such as floods arising from a combination of increased seasonal rainfall and sea-level rise, which could damage continued production of exportable products. Least developed SIDS, which are heavily reliant on tourism, may be particularly vulnerable to climate change[54].

A major aspect of LDC vulnerability to both economic and ecological shocks is the fact that their economies are still in many cases too narrowly based on just a few exports, which themselves are vulnerable to market and other shocks. As the WTO has pointed out, 'irrespective of the size of an LDC, continued efforts to diversify the LDCs' export base need to be made'[55]. This would necessarily include improving LDCs' supply-side capacity to produce and export more diversified products, and shifting away from products that are likely to be vulnerable to climate change.

Trade challenges of SVEs, including SIDS

The trade-related challenges that small vulnerable economies, including most SIDS, face have been identified and articulated by them in, for example, various submissions to the WTO in the context of the Work Programme on Small Economies. Among the most concise but comprehensive of these articulations is a listing of vulnerability characteristics and their implications contained in a submission made in May 2002[56].

Their limited size and constrained natural resource and labour endowments often mean that the domestic markets of SVEs cannot support the location of large-scale industries or the production of goods subject to economies of scale at competitive prices[57]. This means that the range of products produced in SVEs is often limited or products are not priced competitively[58]. SVEs hence often 'show a very high dependence on imports and exports and, consequently, on foreign market conditions' with trade to GDP ratios 'usually much larger than the average' for other developing countries and with exports generally relying 'on a very narrow range of goods and services ... concentrated on the markets of a few countries'[59]. These factors create in many SVEs a high degree of economic instability that both affects and is affected by their vulnerability to climate change.

In a 2006 study conducted by the Commonwealth Secretariat on the performance of 'small states'[60] since 2000, the findings with respect to SVEs' recent economic performance were summarised as follows:

'(a) average GDP growth rates have declined relative to larger low- and middle-income states; (b) income and export volatility remains high; (c) the importance of the service sector (particularly tourism) has risen, while that of agriculture and merchandise exports has declined; (d) remittances and foreign direct investment remain more important to small states than to their larger counterparts; and (e) the debt burden has grown, particularly for Caribbean small states'[61].

More specifically, in the 15 years from 1990 to 2005, SVEs' average growth rate was 3.5 per cent (compared with 4.2 per cent for low- and middle-income countries)[62]. Their general characteristic of being highly open economies deeply reliant on international trade meant that the general economic slowdown in developed countries since late 2001 and the erosion of their trade preferences have adversely affected their own growth rates[63].

The services sector, especially tourism, has become a major source for economic activity in many SVEs[64]. On the other hand, their share of global goods exports has stagnated or risen very slowly (except for those SVEs with 'significant natural resource endowments' such as oil or mineral commodities, whose exports have grown)[65] (see figure 2.2).

A combination of factors stemming from SVEs' (especially SIDS') vulnerabilities to natural disasters, externally driven shocks to their terms of trade (e.g. loss or erosion of trade preferences) and macroeconomic instability, among other things, seem to contribute to SVEs' trade volatility and generally 'less-than-stellar' trade performance in recent years[66].

Some SVEs, especially SIDS, have sought to overcome economic instability and their associated trade-related challenges by pursuing economic diversification strategies. They are 'increasingly becoming knowledge and service-based economies. ... [E]xpanding into eco-tourism and health tourism (health services plus tourism) and in offshore financial services', while others are exploring niche markets for goods[67]. For other SVEs, however, agriculture remains the dominant economic activity to supply both

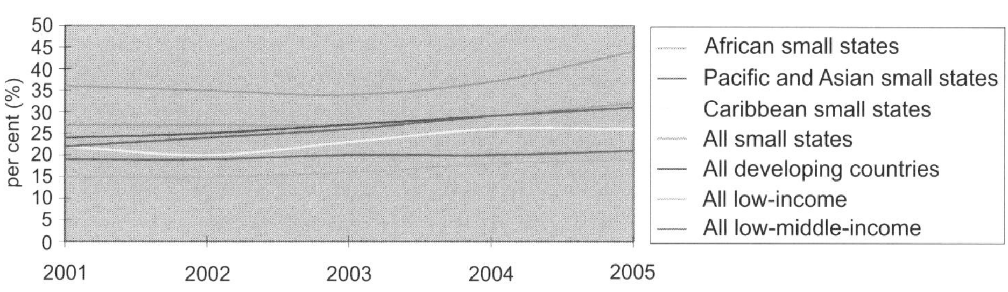

Figure 2.2 SVE export performance relative to other developing countries (goods exports as % of GDP)

Source: Author-generated. Source figures from Commonwealth Small States Review, para.12, table 2.

domestic (with respect to food products) and export markets (with respect to agricultural commodity products such as spices, vegetables, cut flowers and high-quality coffee and cocoa).

Climate change impacts on SVEs' (including SIDS') trade competitiveness

Compared, however, to their larger (in terms of economic size and resilience, land area and natural resource endowments, and population size) developing country counterparts, SVEs, including SIDS, are more vulnerable both economically and ecologically to the impacts of climate change[68]. This means that climate change adaptation is a critical challenge to them (especially for many SIDS), although unfortunately 'they face severe resource and capacity constraints in meeting [that challenge]'[69]. The reliance of their economies in large part on tourism and agriculture as the main economic activities may create development difficulties for them in the face of climate change.

The particular vulnerability of small island developing states to natural disasters such as cyclones, hurricanes, typhoons and other extreme weather events 'continues to pose a formidable challenge to sustainable development for most of them'[70] and the impact 'is exacerbated as a result of the relatively high ratio of coastal area to the land mass and the relatively large population that lives close to the coasts'[71]. Their coastal zones and their fisheries and marine resources are being threatened by 'overexploitation, destructive harvesting, land-based pollution, pollution from ships, coastal development, climate change and invasive alien species'[72]. SIDS' ecosystems are 'small and vulnerable to disruption by climate change or other human activities'[73] with tourism and agriculture projected to be highly impacted[74].

While tourism for many SIDS is a principal economic activity that brings expanded economic opportunities, as a UN report points out, 'tourism and its economic contribution to the economies of small island developing states are threatened by overdevelopment, pollution, loss of diversity, climate change, beach erosion, social and cultural conflict, crime and, more recently, the threat of terrorism'[75]. This could imply that economic diversification efforts to develop the tourism sector may not necessarily provide the developmental benefits hoped for if such threats are not taken into account.

The other main economic activity for many SVEs, including SIDS, agriculture, will also be adversely affected by climate change. Global warming trends leading to temporal and spatial changes in precipitation patterns – including more intense or frequent water-related weather events – are likely to shorten growing seasons in many tropical SVEs, including SIDS. For many of these countries, '[a]rable land, water resources and biodiversity are already under pressure from increases in population on small island states and the unsustainable use of available natural resources. With climate change, negative impacts on agriculture are predicted; coral reefs will be threatened by increased sea surface temperatures and acidification of the oceans; mangroves will be threatened by sea-level rise and an increase in extreme weather events. Water resources are expected to be stressed by changes in precipitation patterns'[76].

Table 2.3 Summary of trade and trade-related climate challenges to development and adaptation for SVEs, including SIDS

Characteristics of current trade profile	*Some projected major climate risks*
• SVEs = <0.16% of global goods trade (<0.1% of non-agricultural market access (NAMA) and <0.4% of agriculture) • SIDS = <0.2% of global goods trade; 0.7% of global services trade (mostly in tourism and financial services) • SVEs, including SIDS, have generally low levels of trade competitiveness, arising from specific circumstances • Generally: o small in land area and population o highly reliant on external trade o limited natural resource base o higher vulnerability to natural disasters o small domestic markets o many are geographically isolated or remote from main markets (island or landlocked) • High levels of export concentration (commodity agriculture, especially sugar and bananas; oil; mineral ores; tourism and financial services); low economic diversification and resilience • Highly reliant on few markets (primarily EU and US)	• Sea-level rise (especially affecting SIDS), affecting physical integrity and infrastructure • Greater frequency of extreme weather events (cyclones, hurricanes, typhoons, storms) • Water stress (arising from de-glaciation and drying up of rivers, desertification, or salination of freshwater lenses and sources) • Overall warming (land and water) resulting in changes in crop growth cycles and crops that can be grown; fish stocks in rivers, lakes and seas • Increased human health hazards, due to changing disease patterns and vectors
Some potential climate change impacts on trade competitiveness, prospects and development	
• Water stress impacts (e.g. drought or the reduction of availability of fresh water due to rainfall changes, saltwater intrusion into freshwater lenses and aquifers etc.) on agricultural productivity (especially small farmer production) may affect current commodity agricultural export performance. • Greater frequency of and vulnerability to more extreme weather events and natural disasters may adversely affect production and export activities due to damage to productive and transport infrastructure. • Sea-level rise may result in damage or loss of coastal infrastructure, including ports and roads (especially in many SIDS), hampering trade transport logistics and increasing transport costs, as well as reducing arable land availability in coastal plains. • Ocean warming may result in coral bleaching and die-offs, adversely affecting near-shore fish stocks and fisheries production, and shifts in ocean currents affecting migratory fish stocks and offshore migratory fisheries productions. • Climate pressure may spur diversification to other crops, products or services for export that are more appropriate to new climate conditions. • Human health impacts may affect labour-force availability and productivity adversely, affecting export performance and economic diversification potential. • 'Tourism could be disrupted through the loss of beaches, coastal inundation, degradation of coastal ecosystems, saline intrusion, damage to critical infrastructures and the bleaching of coral reefs. Physical loss and damage to coasts and infrastructure in SIDS, coupled with the projected milder winters in North America and northern Europe, would threaten the tourism industry by reducing the appeal of the islands as favourable destinations. In addition, the tourism industry may suffer from climate change mitigation measures, such as levies on aviation emissions which would increase the cost of air travel'[77].	

Adapting to climate change: the role of development-oriented trade policy

Linking effective adaptation and sustainable development policy

Adaptation is a process through which societies make themselves better able to cope with an uncertain future. Adapting to climate change entails taking the right measures to reduce the negative effects of climate change (or exploit the positive ones) by making the appropriate adjustments and changes. There are many options and opportunities to adapt. These would range from technological options, such as increased sea defences or flood-proof houses on stilts, to behaviour change at the individual level, such as reducing water use in times of drought and using insecticide-sprayed mosquito nets. Other strategies include early warning systems for extreme events, better water management, improved risk management, various insurance options and biodiversity conservation[78].

The concept of adaptation to climate change is closely linked to the development of adaptive capacity, which refers to changes in processes, practices or structures to moderate or offset potential damages or to take advantage of opportunities associated with changes in climate. The elements involved in the development of national climate adaptive capacity for climate change include, inter alia:[79]

- economic resilience/diversification,
- adequate information on climate risks and impacts,
- human resources,
- financial resources,
- appropriate technology availability,
- natural resources management,
- disaster risk management,
- climate-proofed infrastructure development, and
- existence of policy and administrative mechanisms for integrating adaptation into economic development policy and planning.

All of these elements must be linked by a strong national political determination to carry out climate adaptation as a core component of a country's overall sustainable development strategy[80]. National-level adaptation by developing countries will, in addition to all of the above, also require a strong component of international co-operation as provided for in the UNFCCC from developed countries for support in meeting the costs of adaptation[81]. Finally, effective climate adaptation by developing countries cannot take place unless other climate-adaptive changes are also effected in other areas of international policy, such as trade and intellectual property, to make these fully supportive of developing countries' efforts to undertake sustainable development.

With respect to ensuring the appropriate adaptation of trade policy to climate change impacts and constraints that may arise in a developing country context, especially for smaller developing countries, it will be necessary for policy-makers to move away from the existing approach to economic and trade policy. The trade challenges the smaller developing countries face stem in large part from the confluence of economic and trade liberalisation policy reforms undertaken in the past three decades, which have to a great extent exacerbated these countries' vulnerabilities to both economic volatility and climate change.

Adapting trade-related development policy to climate change

Rethinking trade policy approaches in light of the climate change and development challenges

The climate change and trade policy linkage cannot be approached as a technical issue, but must instead be approached as part and parcel of a broader sustainable development policy framework that would require, inter alia:

- Operationalising sustainable development policy flexibilities and choices for developing countries in the trade and climate change regimes relevant to climate change adaptation,
- Shifting to sustainable, low-carbon, and equitable industrial and agricultural production, diversification and development in developing countries,
- Effecting changes to existing international trade disciplines and commitments (including on intellectual property) vis-à-vis climate-friendly technologies and services in order to provide greater policy flexibilities for developing countries[82],
- Putting in place technology transfer mechanisms consistent with the UNFCCC for climate-friendly technologies at conditions that reflect developing country constraints with respect to access and costs, and which would allow further technology innovation and adaptation by recipient countries,
- Operationalising various options for innovative financing approaches for climate adaptation in developing countries consistent with the provisions of the UNFCCC, and
- Improving trade and climate governance and institution building at the national, regional and international levels (including affected community feedback and accountability mechanisms, South–South regional and cross-regional co-operation on development and climate change).

In the light of their trade-related challenges, and the impacts that climate change is having and will have on the trade competitiveness and economic development prospects of smaller developing countries, as discussed above, it is clear that the key towards effective climate change adaptation by these countries lies in building domestic economies that are resilient, diversified and more productive in sectors that are not as vulnerable to climate change impacts[83].

Further trade liberalisation, in the sense of further increasing levels of tariff openness by smaller developing countries to imports, would not necessarily be the appropriate 'climate-adapted' trade policy response to climate change, for two reasons: (i) LDCs and SVEs, including SIDS, already generally have very open trade regimes, with international trade (both import and export) playing major roles in their economic profile; and (ii) such further liberalisation, when placed in the context of the trade policies and measures being applied by their main trading partners (mostly developed countries), might not result in economic diversification but rather further specialisation and dependence on agricultural and mineral commodity exports[84]. In a sense, with respect to trade policy and climate change adaptation in smaller developing countries, what needs to be done are adjustments to the trade policies and measures of their trading partners more than adjustments to the trade policies and measures of smaller developing countries themselves.

In short, with respect to ensuring that their trade and development policies and measures are climate-adapted, both LDCs and SVES, including SIDS, require broadly:

- Domestic sustainable development policies incorporating both trade and other economic measures designed to build up domestic productive capacity (utilising as much as possible low-carbon processes, technologies and know-how) and to promote economic diversification to other sectors and activities that are not as vulnerable to climate change[85] [86], and
- An enabling and coherent international policy environment in which:
  - o policy space and flexibility for smaller developing countries, in both the trade and climate policy regimes are recognised, reflected and made as broad as possible[87],
  - o new and additional financial flows, in the form of mandatory financing (as under the UNFCCC's Articles 4.3, 4.4. and 4.5), are provided in a manner that is adequate, predictable and responsive to developing countries' sustainable development needs and priorities[88],
  - o the research, development, production and use of climate-friendly technologies take place under conditions where sharing with, transfers to and innovation by developing countries of such technologies effectively happen as a result of a coherent and integrated suite of policy measures, and
  - o external support to smaller developing countries to improve their trade competitiveness is provided[89, 90].

Climate change adaptation of trade-related policy in trade in goods

With respect to trade in goods (both with respect to agricultural and industrial products), climate-adapted trade-related policy-making which is oriented towards sustainable development could imply rethinking smaller developing countries' agricultural development policy and programmes in order to take into account climate change-related impacts on water availability, growing seasons, availability of arable land and continued crop suitability, among other things. This could involve promoting shifts in both crops grown and production processes used to factor in climate change-related stressors – e.g. moving from crops that require a lot of water to ones that do not, utilising water more efficiently, or focusing on new types of crops for both domestic consumption and export, depending on changed climatic conditions locally[91]. Domestic policy shifts may need to be effected – e.g. prioritising agricultural production for food security and industrial production of manufactured goods for domestic consumption over export-oriented production and as a means to lessen vulnerabilities to external trade, economic and environmental shocks[92].

Attention should be paid to the ancillary agricultural and industrial policy and infrastructure shifts (e.g. the use of subsidies, investment promotion measures and support for small- and medium-scale enterprises and, if appropriate, state enterprises) that may need to be made to support diversification away from the current agricultural commodities being primarily exported by smaller developing countries. Adapting to climate change impacts on infrastructure and settlements could include the following: 'providing for the scientific and engineering services required to assess vulnerabilities and define priorities, then retrofitting buildings; integrating adaptation into population and resettlement programmes; improving the planning and permitting processes to guide coastal zone activities, including regulatory adjustments, awareness raising and enforcement; producing design and construction guidelines and applying them in pilot investments'[93].

Climate adaptation will also require that trade-related infrastructure (such as ports, roads, storehouses, refrigeration units etc.) be built or strengthened in preparation for climate impacts (e.g. increased drought and desertification in Africa, increased flooding in Asia, and sea-level rise) that may adversely affect such infrastructure. Adaptation should include, where appropriate, the development of alternative infrastructure if existing ones cannot be climate-adapted.

As with LDCs and SVEs primarily dependent on the agricultural commodities export sector, LDCs and SVEs that are dependent on their oil and other mineral commodity exports will need to ensure that income gains coming from high prices for oil commodities are invested into improving the diversification level of their economies – e.g. by investing in other productive economic sectors and providing for sufficient resources to implement a strategic industrial development policy.

For many SIDS (both LDCs and SVEs), trade in products from their coastal zones and fisheries are major components of their economic and trade profiles. Climate change impacts on these fragile ecosystems, while difficult to combat, could be better addressed through more robust domestic regimes of resource access, control and management.

Climate-adapted agricultural and industrial diversification will also require securing adequate sources of energy to fuel existing production and expansion into new economic activities while, at the same time, not contributing more than is necessary to global greenhouse gas emissions[94]. Often, LDCs and SVEs, including SIDS, are heavily dependent on imports of oil. However, this dependence on imported oil has economic consequences in terms of further compounding the existing economic and natural vulnerabilities of LDCs and SVEs, including SIDS. The development of domestically sourced, clean, renewable, sustainable energy sources and infrastructure is, therefore, an important component for climate-adapted agricultural and industrial diversification.

Climate change adaptation of trade-related policy in trade in services

Economic diversification into the services sectors for smaller developing countries may provide many climate-adaptation and development benefits. Referring to SVEs, the Commonwealth Secretariat pointed out that '[b]y their nature, the service sectors are less vulnerable to the high transport and other infrastructure costs faced by small states, especially the remote ones, and, in contrast to the traditional commodity exports, have robust long-term market prospects'[95]. Services sector expansion and diversification in a manner that is climate adapted will, however, require not only increased domestic public and private sector investment into upgrading the country's human resources (e.g. improvements in education) but also investments in improving and climate-proofing the associated infrastructure for specific services sub-sectors and their various modes of supply.

For example, climate-adapted expansion of tourism services will require strategic adaptation of existing tourism-related infrastructure to take into account climate change impacts (e.g. greater frequency of extreme weather events, loss of terrestrial and marine biodiversity etc.). In the same vein, climate-adapted expansion of other services (such as offshore financial, entertainment services, medical-related services, Internet commerce, software development, tertiary education for foreign students) will require substantial investments into climate-proofing existing and new information and communications technology infrastructure. This will depend, in turn, on ensuring that the associated 'brick and mortar' infrastructure needed for such technology – e.g. data server farms, the energy grid, etc. – are also climate adapted.

Climate change adaptation of the trade-related policies of major trading partners

The major trading partners of smaller developing countries should support climate-adapted agricultural and industrial sector diversification by those countries, including but not limited to the provision of financial and technology transfer assistance for such diversification and immediate action to remove barriers (both tariff and non-tariff) to these countries' exports (especially for those to be developed under the diversification programme). This would mean, for example, that unilateral trade measures being proposed in some developed countries to combat climate change by imposing carbon taxes or duties on the embedded carbon content of imports, insofar as such imports are from smaller developing countries; otherwise such measures be carefully designed so as not to impose disguised restrictions or arbitrary or unjustifiable discrimination against the trade of smaller developing countries.

Developed country subsidies that support unfair competition with the products or production capacity of smaller developing countries, whether in agriculture, industrial goods or even fisheries, should be reduced or eliminated.

Support should also be provided in terms of recognising and allowing for LDCs and SVEs' need for flexibility with respect to the policy shifts and measures for the agricultural and industrial diversification programme. There should be no pressure exerted to require smaller developing countries to provide for additional market access (whether in the agricultural or industrial goods or services sectors) to their trading partners that may not be appropriate in the context of the LDCs' and SVEs' requirements for policy space for climate adaptation of their trade and industrial policy. Any market access opening that smaller developing countries may wish to undertake should be voluntary and flexible to take into account possible changing economic and climatic conditions.

The trade-related actions above should cohere with and enhance the mandatory actions that need to be taken by developed countries under the UNFCCC with respect to the provision of mandatory financing and technology transfers for adaptation, particular in relation to LDCs and SVEs[96], including SIDS. For example, developed country financial support under the UNFCCC should be provided to smaller developing countries in relation to their implementation of specific adaptation measures that have

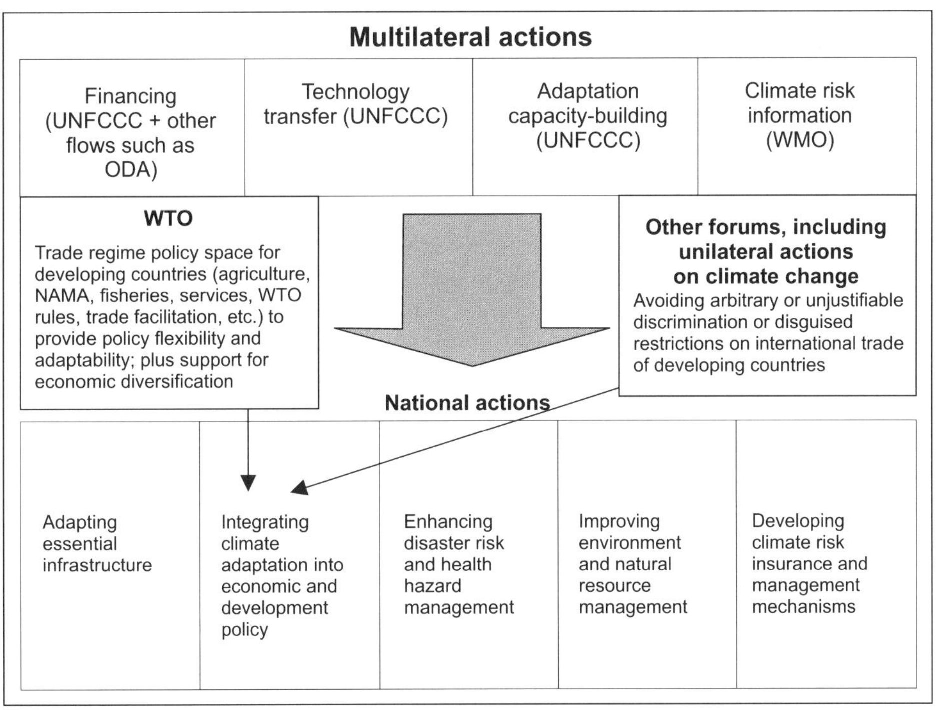

Figure 2.3 Relationship between domestic and international climate change adaptation actions

already been planned by LDCs in the context of their individual National Action Plans for Adaptation (NAPAs) under the UNFCCC.

The interrelationship between domestic climate adaptation trade-policy actions that LDCs and SVEs, including SIDS, can undertake and the supportive actions to be done by their major trading partners as discussed above could thus diagrammed as in figure 2.3.

Finally, in relation to the WTO negotiations, a genuine effort to contribute on trade and climate issues would involve responding effectively to developing country proposals for an integrated and development-oriented approach in determining the negotiated outcomes of the Doha Development Agenda.

Notes

1. See WTO Appellate Body (1998), para.155.
2. For the purposes of this chapter, the category of SVEs will be assumed to also include SIDS.
3. UNCTAD (2007a) Report of the Secretary-General of UNCTAD to UNCTAD XII, para.7 (hereafter UNCTAD XII SG Report).
4. Ibid., para.8. South–South trade in goods is estimated to have increased from US$577 billion in 1995 to US$1.7 trillion in 2005, resulting in a rise of the South–South share of global trade in goods from 11 per cent in 1995 to 15 per cent in 2005. Overall, the share of developing countries in global trade has increased from 29 per cent in 1996 to 34 per cent in 2006. See ibid., paras.15–16.
5. UNCTAD (2006), p. 1–3; see also UN Department of Economic and Social Affairs (2007) World Economic Situation and Prospects, pp.1–10 (hereafter WESP 2007).
6. The pattern of income convergence as a result of growth, according to the UN, seems to be that convergence occurs at the extremes of the income spectrum, where incomes among richer countries tend to converge upwards while incomes among poor countries tend to converge downwards, resulting in greater income disparities between the two groups. See UN DESA (2006) World Economic and Social Survey (WESS) 2006, p. 15.
7. UNDP (2005) Human Development Report (HDR) 2005, p. 37.
8. See IPCC (2007b) Climate Change 2007: Synthesis Report (2007), p.30 (hereafter 4AR Synthesis Report).
9. Ibid.
10. Ibid.
11. Ibid.
12. Ibid.
13. Ibid., p.36.
14. Ibid.
15. Ibid., pp.36–37.
16. Ibid., p.37.
17. Ibid., p.44.
18. Ibid., p.45.

19. Ibid.
20. Ibid., p.46.
21. UNFCCC (2007a), p.8 (hereafter Climate Change Impacts).
22. Ibid., p.6.
23. For a discussion of regional impacts of and vulnerability to climate change in the four developing country regions (Africa, Asia, Latin America and SIDS) see, for example, Climate Change Impacts, pp.18–26.
24. For a ranking of world cities most exposed to coastal flooding arising from climate change, see, for example, Nicholls et al. (2007) (hereafter Nicholls).
25. IPCC 4AR Synthesis Report, p.50.
26. Nicholls, p.3.
27. IPCC 4AR Synthesis Report, p.50.
28. Ibid.
29. Ibid.
30. Climate Change Impacts, p.6.
31. This requirement for co-financing needs to be studied and analysed, with respect to its impacts on access to GEF financing for those developing countries who may not be able to, or may wish not to, have access to co-financing. In many instances, the co-financing requirement has meant that GEF funding is made conditional to co-financing from the World Bank which, with its associated policy conditionalities, may have adverse impacts on the developing country's policy space.
32. UNCTAD (2006a), p.1 (hereafter UNCTAD LDCIII Assessment). The report points out that between 2001 and 2004, only eight out of the 46 LDCs were able to meet or exceed the 7 per cent growth target, with most others growing below that rate, and five LDCs actually showing a decline in GDP.
33. Ibid., p.3.
34. Ibid., p.11.
35. Ibid. LDC growth rates are such that only a few of them 'can be expected to reach the threshold to graduate from LDC status' and 'even supposing that the high growth rates of 2004–2006 continue, only 15 LDCs would have reached the graduation threshold by 2020, including eight which have already reached it. ... To put the overall performance in a comparative perspective, only 11 LDCs were growing at such a pace that their GDP per capita was converging with the average of other developing countries in 2006'. See UNCTAD (2008), p.3 and Box 1 (hereafter LDC Report 2008).
36. The balance-of-payments data for trade in services measure flows relates to modes 1, 2 and 4 of the General Agreement on Trade in Services (GATS). Flows relating to the size of commercial presence are measured through a different framework, that is, Foreign Affiliated Trade in Services statistics (FATS). However, these data are not available for the LDCs.
37. WTO (2008a), para.5 (hereafter LDC Market Access).
38. Ibid., para.7.
39. Ibid., para.10, bullet 2.
40. Ibid., para.10, bullet 3.
41. Ibid., para.12.

42. Ibid. Mineral fuel and oil exports of LDCs in 2005 were valued at US$41.44 billion, while clothing exports were valued at US$12.321 billion. See ibid., para.11, table 4.

43. Ibid.

44. LDC Report 2008, pp.9–11. UNCTAD furthermore pointed out that 'the recent growth surge in LDCs is not generally associated with a structural transition in which the share of manufactures in total output is growing (except for most Asian LDCs). It also indicates the failure to develop productive capacities in LDCs and the weak development of the productive base of their economies, irrespective of strong GDP growth'. Ibid., p.7.

45. LDC Market Access, para.23.

46. Ibid., para.24.

47. Ibid., para.33. The WTO suggests that non-tariff measures in importing countries, such as stringent rules of origin, 'have a direct bearing on the low utilisation of preferences. It is also widely recognised that the limited capacity of LDCs for producing and processing internationally tradable goods is another factor for low utilisation'. Ibid., para.49.

48. Ibid., para.39, Box 2.

49. Ibid.

50. Ibid., para.62. See also WTO (2007); and WTO (2004) (hereafter Supply-Side Constraints).

51. The EU and the US are the LDCs' major export markets, accounting for almost 50 per cent of LDC exports in 2005. China is the third largest export market for LDCs, accounting for 18 per cent.

52. UNFCCC (2007) Climate Change Impacts, pp.18–20.

53. Ibid., pp.21–22.

54. Ibid., p.26, stating that 'Climate change is also likely to have a negative effect on tourism in SIDS, seriously affecting the economy of many small islands. The increasing frequency and severity of extreme weather, sea-level rise and accelerated beach erosion, degradation of coral reefs (including bleaching) and the loss of cultural heritage on the coasts through inundation and flooding are likely to reduce the attractiveness of small island states to tourists...'

55. LDC Market Access, para.13.

56. WTO (2002), paras.4–6.

57. Ibid., para.19.

58. Ibid.

59. Ibid.

60. This phrase is deemed in this paper to be synonymous to SVEs.

61. Lino Briguglio, Bishnodat Persaud and Richard Stern (2006), p.iii. (hereafter 'Commonwealth Small States Review').

62. Ibid., para.7.

63. Ibid.

64. As the Commonwealth study points out: 'The average share of tourism in export receipts for the small states in Africa now stands at about 20 per cent, while the figures for Asia and the Caribbean are 45 per cent and 30 per cent, respectively. The comparable figure for all low- and middle-income developing countries is about 7.5 per cent'. Ibid., para.11.

65. Ibid., para.12.

66. Ibid., para.10. In particular, in relation to the impacts of climate change, '[t]he impact of natural disasters has tended to be higher in those countries that are more dependent on agricultural production'.

67. Ibid., para.46. See also UN (2004) Report of the Secretary-General, para.5. (hereafter 'SIDS POA Review').

68. The SE Literature Review, para. 16, notes that '[T]he impact in terms of per capita costs and per unit of area damage of strong hurricanes and disruptive earthquakes are much more severe in smaller countries'.

69. Commonwealth Small States Review, para.98.

70. SIDS POA Review, para.39.

71. Commonwealth Small States Review, para.96.

72. SIDS POA Review, para.46.

73. Ibid., para.55.

74. UNFCCC (2007b), para.9. (hereafter 'UNFCCC SIDS Adaptation')

75. SIDS POA Review, para.14.

76. UNFCCC SIDS Adaptation, para.8. On many SIDS, prime agricultural land is located on the coastal plains, which are threatened and likely to be affected by sea-level rise. In addition to this, reductions in the availability of freshwater supplies (whether from rainfall or freshwater lenses and aquifers), aside from the impacts of weather events that affect growing cycles, could further damage agricultural production, leading to major economic damage and affecting food security.

77. UNFCCC (2005), p.23.

78. Climate Change Impacts, p.10.

79. See e.g. IPCC 4AR Synthesis Report, p.56, stating that 'The capacity to adapt and mitigate is dependent on socio-economic and environmental circumstances and the availability of information and technology'.

80. Ibid., stating that 'Many adaptation actions have multiple drivers, such as economic development and poverty alleviation, and are embedded within broader development, sectoral, regional and local planning initiatives, such as water resources planning, coastal defence and disaster risk reduction strategies'.

81. See UNFCCC, Article 4.4. For example, the IPCC has pointed out that the cost of adaptation to climate change for Africa 'could amount to at least 5 to 10 per cent of GDP'. See IPCC 4AR Synthesis Report, p.50.

82. For example, would compulsory licensing arrangements under the TRIPS Agreement for climate adaptation technology similar to that provided for in the TRIPS and Public Health Decision of 30 August 2003 and the 2005 Amendment to the TRIPS Agreement with respect to pharmaceutical products be useful for purposes of enhancing technology transfers of such technology to developing countries? See http://www.wto.org/english/tratop_e/trips_e/implem_para6_e.htm [accessed 6 March 2009] and http://www.wto.org/english/tratop_e/trips_e/wtl641_e.htm [accessed 6 March 2009] for the texts of this Decision and Amendment. Another question is how LDCs can maximise the current extension given to them in 2005 allowing them an additional transition period under Article 66.1 of the TRIPS Agreement until 1 July 2013 for them to put in place TRIPS Agreement implementing measures to provide protection for trademarks, copyright, patents and other intellectual

property (see http://www.wto.org/english/news_e/pres05_e/pr424_e.htm [accessed 6 March 2009] for the text of this decision). That is, could this additional transition period be used by LDCs to spur technology transfer of climate-friendly technologies using measures that may be later on TRIPS-inconsistent?

83. See e.g. LDC Report 2008, p.II, stressing that 'LDCs need to improve agricultural productivity and diversify their economies to create non-agricultural employment opportunities'.

84. See e.g. Commonwealth Small States Review, para.35, pointing out that '"effective market access" created through the removal of internal barriers to trade does not necessarily result in effective access for small states' exports into developed country markets'.

85. It should also be noted that under Annex F (Special and Differential Treatment) of the WTO's 2005 Hong Kong Ministerial Declaration, LDCs are allowed to maintain measures that are inconsistent with the WTO's Agreement on Trade-Related Investment Measures (TRIMS), such as performance requirements, local content requirements etc., up to 2020. See http://www.wto.org/english/thewto_e/minist_e/min05_e/final_annex_e.htm [accessed 6 March 2009]

86. In this context, information on climate change impacts needs to be translated into 'language and timescales relevant to policy-makers', and research on 'the potential impacts of climate change needs in-country support to enable information to be improved and passed on to policy-makers' and those implementing such policies on the ground. See e.g. Saleemul Huq et al (2003), p.36.

87. This issue of having trade-related policy space is clearly reflected in various LDC negotiating proposals and submissions in the context of the WTO's Doha Development Agenda negotiations. SVEs, including SIDS, have also highlighted their need for trade-related policy space. For both LDCs and SVEs, including SIDS, the following comments from the Commonwealth Secretariat with respect to policy space and small states' trade and development are of great relevance: 'The request for appropriate "policy space" is associated with the realisation that small states' strict adherence to WTO disciplines may, to some extent, limit their room for policy manoeuvre. Although small states recognise the potential gains from trade liberalisation, some compromise needs to be sought to introduce a degree of flexibility, such as for example derogations from certain subsidies disciplines to allow small states to provide incentives for attracting FDI to compensate for the high cost of doing business in small economies and appropriate provisions negotiated on Special Products (SP) and Special Safeguard mechanisms (SSM) for the development of their agricultural sector. The challenge will be to introduce some flexibility while at the same time not hollowing out the rule based trading system, which has significant potential benefits for developing countries'. See Commonwealth Small States Review, para.35. In the context of climate regime policy space, the implementation of the existing balance of commitments under the UNFCCC needs to be enhanced and strengthened, and unilateral initiatives or measures that may be undertaken by the trading partners (mainly developed countries) of the LDCs and SVEs, including SIDS, 'to combat climate change ... should not constitute a means of arbitrary or unjustifiable discrimination or a disguised restriction on international trade' (UNFCCC Article 3.5) especially with respect to the export products of LDCs and SVEs, including SIDS.

88. The focus in terms of climate financing (including with respect to adaptation) should be with respect to fulfilment by developed countries of their financing commitments under the UNFCCC, rather than in terms of increasing donor-driven and voluntary aid flows.

89. For SVEs, including SIDS, the Commonwealth suggested that '[f]ive types of external support were identified as being particularly helpful – agreement on transition periods, provision of financial assistance, action to remove barriers to small states' exports (including agricultural exports), support that addressed capacity and vulnerability problems, and support for participation in the WTO and international trade discussions'. These are also equally applicable to LDCs. See Commonwealth Small States Review, para.26.

90. External official development assistance (ODA) flows to LDCs and SVEs, including SIDS, leave much to be desired in terms of both quantity and quality. In terms of quantity, the LDC Report 2008 (p.IV), for example, has pointed out that aid intended to build productive sectors and economic infrastructure in LDCs 'has continued to receive less priority ...[at] just 25 per cent of total ODA commitments to LDCs in 2006' and that such aid, in fact, has been on a decreasing trend. In terms of developmental quality, there is little indication that the current aid system has become more responsive to recipients' needs and requirements – a critique that has also been raised in discussions within the OECD's Development Assistance Committee. Under current arrangements, aid recipients' (many of them LDCs and SVEs) dependence on donors will continue indefinitely, with little prospect of their exiting from aid altogether. Hence there should be a significant shift in the framework for the delivery of aid towards one where developing countries are supported and enabled to exit from aid dependence, which would be the true test of aid effectiveness. The ultimate goal of all aid should be to enable recipients to become aid independent within a reasonable time frame, and this should be the context for evaluating aid effectiveness.

91. Examples of climate change adaptation with respect to biodiversity, land and water resources relevant to discussions relating to diversification of the agriculture sector were identified in a UNFCCC as follows: 'Measures to address the impacts of climate change on biodiversity and land degradation include: creating land-use plans and corresponding enforcement strategies; rainwater harvesting, water-demand management, provision of water storage and water-efficient household appliances; flood-risk analysis with land zoning and flood mitigation actions; strengthening of institutional capacity to enforce land zoning restrictions, including the application of beach setbacks for construction; the use of land-use models in order to make farming more efficient and less destructive to the environment; and training fishermen and women in sustainable fishing practices. ... Measures that have been suggested to respond to projected decreases in water resources include: incentives to encourage the use of water-saving devices; selecting appropriate drought tolerant vegetation; establishing river buffer zones to enhance the resilience of the river and catchment area; updating national water policies, improving water resources management; revising building codes to increase opportunities for rainwater catchment and storage; preparing water resource master plans for islands; and assessing and improving the water supply system'. UNFCCC SIDS Adaptation, paras.74 and 76.

92. Taking account of new climatic conditions, questions that domestic policy-makers in LDCs and SVEs, including SIDS, will need to ask could include the following: (i) what crops should their agricultural sector focus on to enhance food security and economic gains?; (ii) what manufactured goods could their existing industrial sector produce that can be used to lay the foundations for a broader-based domestic industrial development process and lessen exposure to external vulnerabilities?; (iii) what investments in public infrastructure would be required to support climate adaptation?; (iv) how can financial resources be generated both domestically and from external sources to finance broad-based development-oriented climate adaptation?; (v) how can existing international arrangements and commitments,

including flexibilities therein, be maximised to support the acquisition and appropriate innovation of technologies to support climate-adapted development?

93. UNFCCC SIDS Adaptation, para.75.

94. LDCs and SVEs, including SIDS, are currently among the lowest per capita greenhouse gas emitters, with an average of less than 1 ton of carbon dioxide equivalent per year.

95. Commonwealth Small States Review, para.129.

96. It should be noted, however, that the term 'vulnerable' has specific political and technical connotations in the UNFCCC context and is usually qualified with 'most' or 'particularly'. References to this criterion are found all over the UNFCCC (preambular para.19), Article 3.2 for adaptation, Article 4.4 on costs for adaptation, the listing in Article 4.8, as differentiated in the use of the term 'vulnerable' in UNFCCC Article 4.10 and Article 12.8 of the Kyoto Protocol. This distinction in the use of the term 'vulnerable' between the two fora of discussions (climate change and WTO) is important to bear in mind, in particular when referring to adaptation to the adverse effects of climate change.

3

Trade and Transport

3.1 International Transport, Tourism and Services

Keith Nurse, University of West Indies, Barbados

The international transport sector, including shipping and aviation, is of a particular nature, as reducing their greenhouse gas emissions does not fall directly within the jurisdiction of any country. Traditionally, the emissions have been out of sight and out of mind, and so regulation in international transport has been lagging. Discussions are now ongoing within the UNFCCC and the International Maritime Organization (IMO), and voluntary approaches are being considered within the International Civil Aviation Organization (ICAO).

Recent studies by the IMO show that shipping contributes much more carbon dioxide than previously thought. The worldwide fleet of 90,000 ships transports 90 per cent of the world's traded goods, and contributes 2.7 per cent of carbon dioxide emissions, up from a previous estimate of 1.8 per cent. Shipping has grown by 3 per cent annually on average over the last three decades, and shipping emissions are projected to grow by more than 70 per cent by 2020, as global trade expands. Aviation emissions contribute between 2–9 per cent of global emissions, depending on how they are calculated, and are also growing rapidly.

The regulation of emissions from international transportation would potentially mean raising costs for moving goods and people around the globe, with implications for international trade. Developing countries situated in remote locations, such as SIDS, and developing countries with large trade exposure, would be particularly affected by higher bunker fuel prices. Certain service sectors such as tourism would likely be particularly affected by new emissions cuts in the international transport sector.

Current discussions on international levies on bunker fuels, or an international trading scheme in this area, need to consider this fact, and the principle of common but differentiated responsibilities taken into account. At the same time, the potential costs of climate regulation need to be seen in relation to their benefits, and other costs, such as fluctuations in energy costs unrelated to climate regulation.

Small island developing states: challenges and opportunities

Small island developing states (SIDS) are not large contributors to the problem of climate change, but are estimated to be impacted the most. In terms of definitions, the

Barbados Plan of Action coined SIDS primarily as 'limited in size, hav[ing] vulnerable economies [and are] dependent upon narrow bases and on international trade, without having the means to influence that trade' (Annex 1, Part 1 [IV], Barbados Plan of Action)[1]. While the key climate issue for SIDS is adaptation, they are highly dependent on the tourism and travel industries, which are considered to be major emitters – and so are vulnerable to the effects of international climate mitigation policies in these sectors.

Adaptation costs can be devastating. This predicament is exacerbated by the fact that many SIDS are dependent upon mono-crop agricultural production and export and tourism for foreign exchange earnings, employment and contributions to GDP. These countries are also highly dependent on the importation of food and energy for domestic consumption and for the tourism sector.

SIDS, which are generally long-haul destinations from key source markets like North America and Europe, have raised concerns regarding the potential adverse impact of prospective climate regulation of the air travel and shipping sectors and consumer preferences shifting in favour of short-haul destinations.

At the international level, developing countries, and particularly SIDS, recognise their state of vulnerability to climate change and therefore urge a focus on adaptation and support from those parties responsible for climate change, which need to take a lead on mitigation. At the same time, SIDS are also advancing a proactive agenda looking at adaptation and mitigation in tandem, urging the development, dissemination and transfer of efficient energy technologies that can assist developing countries in mitigating the effects of climate change. SIDS have also acknowledged their responsibility to collect data on the effects and implications of climate change and sea-level rise, to improve public understanding of the issue, to promote more efficient energy use and to formulate their own comprehensive adjustment and mitigation policies to be able to cope with and respond to climate change. SIDS also co-operate at the regional level to respond to the climate change challenge, and work with the international aviation and cruise line industries, as well as NGOs that seek to promote sustainable tourism, to improve their climate profiles.

In order to move from a position of vulnerability and dependence to one of resilience, policy tools within the international trade arena can be used to boost the capacity of SIDS. The services sector, and in particular tourism, represent a genuine opportunity for such countries to expand their economic activity while earning foreign currency.

In addition, SIDS can seek to liberalise trade in energy-efficient goods in a bid to decrease their collective carbon footprint. This policy could include both tax incentives and zero-tariff measures for the import of environmentally friendly products. The trade arena could also facilitate the transference of technologies that contribute to the development of capacity among service providers. This can indeed be particularly useful as practitioners from SIDS within the tourism industry (and other industries as well) sometimes find the cost of technological devices to be prohibitive.

Technology transfer can also be important for environmentally friendly technologies for local industries, and meteorological technology to inform tourists and industry officials of impending bad weather, especially severe natural hazards, enabling officials to take pre-emptive action to ensure the safety of citizens and tourists.

Perhaps one of the most direct and legally binding approaches that a group of nations can adopt is to sign a trade agreement that addresses issues closely related to climate change. An example can be found in the recently initialled Economic Partnership Agreement (EPA) between the Caribbean Community and the Dominican Republic (CARIFORUM) and the European Union (EU).

This section examines the challenges and opportunities for SIDS with regard to climate change and tourism, travel and related services, as well as relevant policy tools and instruments in this area.

Climate impacts on SIDS and adaptation costs

The predictions of scientists suggest that SIDS are highly vulnerable to the impacts of global warming, particularly in terms of sea-level rise, temperature rises, rainfall changes, coral bleaching and increased storm frequency. SIDS are in fact considered 'hotspots' for climate impacts, and are one of the groups of countries estimated to be most affected by climate change. For example:

> Sea-level rise will exacerbate inundation, erosion and other coastal hazards, threaten vital infrastructure, settlements and facilities, and thus compromise the socio-economic well-being of island communities and states.

Warming seas threaten the livelihood of commercial and artisanal fisheries and coral reefs. This, by extension, has the potential to result in widespread unemployment of fishermen and of tourism-related service providers (e.g. scuba tourism), whose incomes depend upon the existence of healthy coral reefs. To add to this, if climate change does result in changing rainfall distribution patterns, then many SIDS will be forced to find new and innovative ways to establish a consistent and reliable water supply. The absence of a consistent water supply can also lead to severe declines in agricultural production (subsistence and commercial), thus threatening food security within these island states along with the competitiveness of the tourism sector.

The challenges of climate change are no longer futuristic. Kiribati, with a population of 100,000, is estimated to become 'the first sovereign victim of man-made climate change'. The adaptation costs can be quite devastating. Their predicament is exacerbated by the fact that many SIDS depend on mono-crop agricultural production and exports, as well as tourism and its associated services for foreign exchange earnings, employment and contributions to GDP. These countries are also highly dependent on the importation of food and energy for domestic consumption and for the tourism sector.

Tourism – a double-edged sword

The travel and tourism sector is the key economic sector for SIDS in terms of earnings and jobs. Many SIDS are highly dependent upon revenue earned from tourist arrivals and through tourist-related activities, while tourism earnings account for a significant share of the foreign exchange earnings in most SIDS. With regards to the Caribbean, travel and tourism accounts for 14.8 per cent of GDP, 12.9 per cent of employment and 14.6 per cent of total exports. Oceania has a similar economic profile, with GDP shares of 11.7 per cent, employment 12.4 per cent and exports of 16.9 per cent. However, for both regions ten-year forecasts (2018) by the World Travel and Tourism Council (2008) suggest declining contributions to GDP and employment, but not to exports.

The air travel sector and the cruise ship industry provide key services to the tourism sector in small island states, which are generally long-haul destinations from key source markets like North America and Europe. Notwithstanding this, the travel sector is considered a major contributor to green house gases, and there is a potential threat from tourism source countries in terms of taxation schemes and consumer movements that may deter holidaymakers from long-haul travel. The intersection of these factors makes for a critical scenario for SIDS in the evolving context of climate change and trade in international services.

In addition, there are concerns among tourism authorities in SIDS that many tourists are choosing to remain within their own country or region during their vacation period. This has prompted governments and companies to adopt and promote environmentally friendly charges, levies and technologies, some of which have caused the cost of travel and transportation to increase. Increased travel and transportation costs will likely have adverse effects on travel and tourism to SIDS. Table 3.1, below,

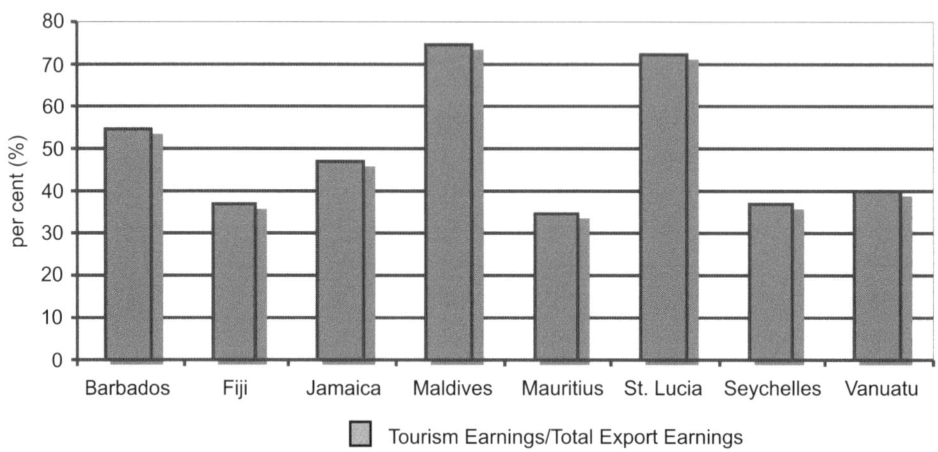

Figure 3.1 Tourist receipts as a share of total export earnings, selected SIDS (2004)
Source: Commonwealth Secretariat.

Table 3.1 Hypothetical results of alternative growth rates in international tourism flows to the Caribbean region 2000–2050 (millions arrivals)

Flows	*Growth rate (% pa)*	*2000 (the base year for the flows*	*2020*	*2030*	*2040*	*2050*
Base case exploitation						
N America to Caribbean	3.0	8	14	19	26	35
Half point growth rate reduction						
N America to Caribbean	2.5	8	13	17	21	27
Full point growth rate reduction						
N America to Caribbean	2.0	8	12	14	18	22

Source: Travel Research International.

provides three scenarios in terms of alternative growth rates for arrivals from North America over the period 2000 to 2050. It suggests that the Caribbean could have a drop-off in arrivals from North America of some 13 million when the best and worst cases are compared.

The direct costs of climate change, beyond their effect on the development of the tourism sector, are also expected to be very high in SIDS. Recent estimates of the impact of climate change on the Caribbean region paint a dismal picture. Key findings of a study[2] on the cost of inaction concludes that:

- The costs of inaction will amount to 22 per cent of gross domestic product (GDP) for the Caribbean as a whole by 2100,
- The costs of inaction will reach an astonishing 75 per cent or more of GDP by 2100 in Dominica, Grenada, Haiti, St Kitts and Nevis and Turks and Caicos,
- The Caribbean's largest island, Cuba, faces a nearly 13 per cent economic hit by mid-century, and a 27 per cent loss by 2100, unless there is swift action to address climate change,
- Losses from inaction would be less severe, but still significant in Puerto Rico, reaching nearly 3 per cent by 2050 and 6 per cent by the end of the century, and
- The nation of Colombia, with its long Caribbean coastline, faces permanent flooding of 1,900 square miles in low-lying coastal areas, affecting 1.4 million people.

Policy tools and instruments

There are a number of mechanisms to combat climate change and promote sustainable tourism, although the two seldom, nor automatically, go hand in hand.

Inter-governmental action

Developed countries are under obligation to lower their emissions, as codified under the UN Framework Convention on Climate Change (UNFCCC), with most developed countries having taken on additional, legally binding targets under its Kyoto Protocol. Many of these countries have also adopted other strategies and measures geared towards lowering their carbon footprint, investing in research geared towards increasing energy efficiency, fuel economy and use of renewable sources of energy[3].

Developing countries, and particularly SIDS, recognise their state of vulnerability to climate change and therefore urge the development, dissemination and transfer of efficient energy technologies that can assist them in mitigating the effects of climate change. As such, developing countries recognise the importance of developed countries 'taking the lead' in this regard[4]. SIDS have also acknowledged their responsibility to collect data on the effects and implications of climate change and sea-level rise, to improve public understanding of the issue, to promote more efficient energy use and to formulate their own comprehensive adjustment and mitigation policies to be able to cope with and respond to climate change. In sum therefore, developed and developing nations tend to respond to the threat of climate change in a way that is consistent with international consensus (as expressed through the UNFCCC), where nations take measures to protect the earth's ecological system through policies and instruments that reflect their common but differentiated responsibility[5].

MEAs and the tourism industry

The UNFCCC does not address the tourism industry directly. There are, however, many multilateral environmental agreements (MEAs) that can be said to have the potential to serve the interests of the tourism industry, particularly in SIDS. These include, for example, the Convention on International Trade in Endangered Species of Wild Flora and Fauna of 1973 (CITES), which serves to protect endangered species and thus can preserve species that are of importance for tourism. Other examples include the Convention for the Protection of the World Cultural and Natural Heritage and the Protocol Concerning Specially Protected Areas and Wildlife to the Convention for the Protection and Development of the Marine Environment of the Wider Caribbean Region. The latter includes objectives that not only help to strengthen environmental protection in SIDS, but also helps to safeguard the future of eco-tourism in these regions.

One should also note that the UN Global Conference on the Sustainable Development of SIDS (1994) adopted the Barbados Programme of Action on the Sustainable Development of SIDS (BPOA) – which addresses, inter alia, climate change, natural disasters and tourism. Such resolutions can also be used as avenues for action by the international community.

Industry initiatives

The International Air Transportation Association (IATA) estimates that aviation is responsible for 2 per cent of carbon dioxide emissions and 12 per cent of emissions from all transport sources. IATA has adopted a four-pronged approach to reducing emissions from the industry[6]. To begin, IATA's strategy focuses upon technological advancements through research and development (R&D) in order to improve the fuel efficiency of aircrafts. Secondly, aircraft operators are encouraged to improve their internal operations and to ascribe to high environmental standards. Further, IATA urges governments to correct infrastructural inefficiencies (particularly in airports) and to take steps to make global airspace more free for air transport carriers. Finally, economic measures are suggested as a means to boost R&D and to improve environmentally friendly practices within the industry. As such, tax credits and direct funding for eco-friendly measures, as well as an open emissions trading scheme are asserted as possible mechanisms that can be utilised to decrease the aviation industry's contribution to climate change.

As the global discourse on climate change continues to grow, more and more tourists may decide to stay at home or within their region during their vacation period to lower their own carbon footprint. If the aforementioned measures suggested by IATA were successful in lowering emissions from the aviation industry, this might encourage more tourists to fly longer distances to SIDS for their holidays.

The cruise ship industry has also been taking steps to decrease its contribution to environmental degradation (particularly the destruction of coral reefs) and to global warming. One such initiative was announced in January 2008 when stakeholders from the cruise industry, the Mexican government and Conservation International (CI) unveiled a plan to 'protect coral reefs and other ecosystems in Cozumel [Mexico], the world's most-visited cruise destination'[7]. This arrangement is part of the Mesoamerican Reef Tourism Initiative (MARTI) and it aims to facilitate co-operation between industry and government officials, as well as representatives from civil society. The agreement aims to raise environmental awareness on and off of cruise ships, to improve management of waste and infrastructure, to promote increased protection of the reef system, in addition to enhanced enforcement of existing laws and regulations. The MARTI was also extended to Belize in May 2008 when the cruise industry, government and NGOs signed a Declaration of Commitment to sustainable tourism practices[8].

Individual cruise liners have also taken steps to lower their emissions through, inter alia, partnering with institutions engaged in research on the subject. Such initiatives are yet to be implemented in SIDS, but they point to the new directions that the cruise ship industry can adopt to reduce emissions and bolster consumer confidence in the climate change agenda.

Involvement by non-governmental organisations (NGOs)

International and national NGOs have helped to shape not only the discourse, but also the social movement related to the issue of climate change. They have been able to lobby governments, corporations and other international organisations to take action to combat climate change, and have engaged in widespread public awareness and education campaigns to spread knowledge about climate change. They have also played an integral role in outlining the effects of climate change on the environment, the global economy and on specific industries (like the tourism industry).

Additionally, the World Wide Fund for Nature (WWF) makes a very interesting link between conservation and tourism. In so doing, the organisation promotes arctic tourism as an avenue to alert persons to the effects of climate change. The WWF also asserts that tourism and conservation are compatible and as such seeks to give tourists useful hints on how they can enjoy their vacation in an environmentally friendly way[9]. Thus this international NGO utilises the global climate agenda both to mitigate the effects of climate change and to promote tourism.

The role of regional organisations

Many nations, including SIDS, have chosen to respond to the threat of climate change through regional institutions. As such, they gain wider consensus on this global issue and regional institutions often are more cost-effective (as the states involved would be able to share the cost involved).

For example, the Caribbean Community Climate Change Centre (CCCCC) seeks to act as:

> 'A centre of excellence for the development of policy, technical research and the mobilisation of financial and other resources to address climate change and related matters within the community'[10].

The CCCCC delivers information to member states of the Caribbean Community (CARICOM), assisting them in making policy geared towards adopting measures to mitigate and adapt to the effects of climate change. The organisation also serves as an information node, to facilitate networking between the public and private sector, knowledge-based institutions and individuals. Furthermore, the CCCCC engages in forecasts and analyses of the impact of climate change on the environment (especially coral reefs and other coastal and marine resources), the economies and specific industries, such as tourism in the Caribbean region. This approach to addressing the threat that climate change poses to the global environment helps to build international solidarity around the issue and also gives region-specific analysis of the problems that a warmer earth will pose to individual countries. Such a perspective should indeed be helpful when probing the effects that climate change might have on tourism in SIDS in different atmospheric regions of the world.

From vulnerability to resilience – the role of trade policy

In order to move from a position of vulnerability and dependence to one of resilience, policy tools within the international trade arena can be used to boost the capacity of tourism services providers within SIDS. The services sector, and in particular tourism, represents a genuine opportunity for such countries to expand their economic activity while earning foreign currency.

Countries can also liberalise trade in energy-efficient goods in a bid to decrease their collective carbon footprint. Such a policy instrument can include tax incentives or zero-tariff measures for environmentally friendly products. Notwithstanding this, in seeking to adopt trade policies and instruments to combat climate change, countries need to ensure that such initiatives do not become or constitute a barrier to international trade. This principle is embraced by the UNFCCC, which states that 'measures taken to combat climate change, including unilateral ones, should not constitute a means of arbitrary or unjustifiable discrimination or a disguised restriction on international trade'[11].

Countries could also take measures to encourage climate-friendly imports. Based on existing WTO case law, it may be prudent to assert that countries seeking to adopt policy measures and instruments to combat climate change should not try to influence the way merchandise is processed, harvested or produced in other countries through unilateral trade measures. Instead, international consensus should be sought, particularly through MEAs, to formulate innovative mechanisms to conserve the climate as an exhaustible natural resource without creating barriers to international trade.

The trade arena can also be utilised to facilitate the transfer of technologies that can contribute to the development of capacity among service providers. This can indeed be particularly useful as practitioners from SIDS within the tourism industry (and other industries as well) may sometimes find the cost of technological devices to be prohibitive. Moreover, a lack of access to technology can also act as a technical barrier, preventing service providers from operating at international standards. As a result, trade negotiations (especially as they pertain to North–South trading arrangements) often feature facilities for the transfer of technology. This is a particularly useful policy instrument for governments seeking a cost-effective way to acquire more environmentally friendly technologies for local industries. Furthermore, the transfer of meteorological technology can help to inform tourists and industry officials of impending bad weather, especially severe natural hazards, to enable officials to take pre-emptive action to ensure the safety of citizens and tourists. Such technology has the potential to dramatically improve climate forecasts, which can enhance the ability of governments and tourism officials to warn prospective tourists of potentially dangerous and extreme weather patterns and events brought about by climate change.

Perhaps one of the most direct and legally-binding approaches that a group of nations can adopt would be to sign a trade agreement that also addresses climate change issues. The recent Economic Partnership Agreement (EPA) between the Caribbean Community and the Dominican Republic (CARIFORUM) and the European Union (EU) provides such an example. The EPA represents a comprehensive trading arrangement between an archipelago of SIDS and a group of developed nations. However, in addition to expressing the overall objective of trade for sustainable development, the agreement contains a chapter on the environment. In it, the parties to the agreement reaffirm their commitment to:

> 'Conserve, protect and improve the environment, including through multilateral and regional environmental agreements to which they are parties' and to 'promoting the development of international trade in such a way as to ensure sustainable and sound management of the environment, in accordance [with]...international conventions to which they are party and with due regard to their respective levels of development'[12].

Thus, while the chapter (or the entire agreement) does not specifically mention or target climate change, it provides the scope and framework for measures to be adopted in the interest of tackling climate change. Additionally, the EPA contains provisions pertaining to co-operation on issues pertaining to the environment, through technical assistance, trade in natural resources and through public education campaigns to foster trade in environmental goods and services[13].

In tandem, it should also be noted that the EPA seeks to foster trade in goods and services, which

> 'Parties [to the agreement] consider to be beneficial to the environment. Such products may include environmental technologies, renewable – and energy-efficient – products and services and eco-labelled goods'[14].

Small island developing states thus need to carefully evaluate the opportunities and challenges presented within the trade realm with regard to climate change, adaption, tourism, travel and their services industries.

3.2 Food Miles Debate

Hasit Shah, Chair, Fresh Produce Exporters Association of Kenya (FPEAK)

The world trade in fresh fruits, fruit products, vegetables and vegetable products has more than doubled over the past two decades between 1982–1984 and 2002–2004 (Mildon, 2007). Developing countries export a third of that total trade. The social and economic benefits of this trade to developing nations – including least developed countries, small vulnerable economies and small island developing states – are crucial as it generates income and jobs.

The first experiment with a crude form of labelling – airplane stickers in supermarkets to indicate fresh produce that had been airfreighted – ended up hurting some of the poorest and most vulnerable countries. These developing countries had managed to capture high-value niche markets in developed countries by airfreighting fresh produce during the northern winter. The stickers singled out just one part of the carbon footprint, namely transport, ignoring other parts of the process. Overall, the exporters operating in warmer climates often produced such goods with lower carbon emissions as compared to their counterparts in developed countries, which produce out-of-season vegetables in a highly mechanised fashion in greenhouses using large amounts of carbon-intensive fertiliser input. Under a differently designed scheme, the small developing country producers may have been the ones to benefit.

The role of voluntary carbon-labelling schemes is likely to grow in the future, providing consumers with the option of decreasing their personal carbon footprints. Given this context, the debate on food miles needs to be expanded to not just include road and sea transport, but to look at the total carbon emissions of a product through the supply chain, using life-cycle analysis, and evaluate how to reduce emissions at each stage of the chain to achieve a carbon neutral rating.

Overall, carbon-labelling schemes provide opportunities as well as challenges for developing countries. Any future carbon schemes would need to balance the need for accurate and useful data with the need to be simple, transparent and to involve sufficiently low transaction costs to include small countries and players. Labelling schemes provide opportunities for positive product differentiation and market opportunities. On the other hand, many producers are concerned that labelling and standards become barriers to market access. They see the rise in such 'non-tariff barriers' as potential obstacles to market entry and as a vehicle for green protectionism.

Many developing countries feel they are not represented and their voice is not heard in the development of private-sector labelling schemes. They have limited room to manoeuvre, given that the international trade rules governing standards and technical regulations, namely the World Trade Organization (WTO) Agreement on Technical Barriers to Trade and the WTO Agreement on Sanitary and Phytosanitary Standards (SPS) essentially binds member countries, not private organisations. However, there is now an ongoing debate in the issue of private-sector standards within the WTO SPS Committee, questioning this notion.

Global trade in fresh produce

The value of the global trade in fruit and vegetables is estimated to be around US$45 billion, with over 73 million tons moving around the world (Legge et al., 2006). Developing countries export a third of the total, of which eight countries account for two-thirds of the total. These are countries that have aggressively developed their agricultural sectors, with high levels of technology transfer and competent Sanitary and Phytosanitary (SPS) protocols that have allowed them access to lucrative markets. Some of these countries, like Chile, Ecuador and Mexico, have access to the US market, while China has access to the EU and Japan. They also have favourable climates, cheap labour and well-developed logistics and transport chains.

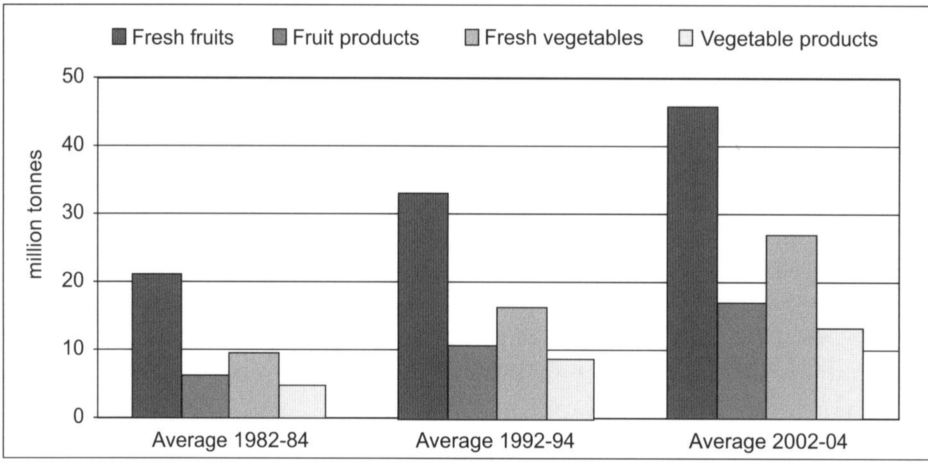

Figure 3.2 World fruit and vegetable exports (million tonnes)
Source: Mildon, 2007.

As the global population continues to increase to a projected eight billion by 2025 and over nine billion by 2050 (UN), the value and amount of trade will continue to grow substantially. Per capita consumption is increasing, possibly due to a growing awareness of the nutritional benefits associated with fresh produce and as populations become affluent and are exposed to a range of fruits and vegetables that were not part of the daily palate in the past.

The social and economic benefits of this trade to developing nations are crucial as it generates income and jobs. According to MacGregor and Vorley (2006), over one million people in rural Africa are supported by fresh fruit and vegetable (FFV) exports to the UK alone, and the trade returns an estimated 200 million UK pounds (£) to the rural economy.

Carbon labelling

Labelling of FFV with certain information has become a legal requirement throughout the world. Basic information about contents, weight, 'best before' dates, nutritional values and origin are standard. Over the years, retailers have added on more information to the label for the benefit of the consumers including traceability, recipes and different accreditations such as 'Fair Trade' and 'Organic'. Among the newer labels are those related to the carbon emissions associated with FFV. The greenhouse gas emissions include those related to production, the various modes of transport, refrigeration and the cold chain necessary for agricultural trade.

Initial labelling of imported produce that was air flown was proposed by two leading food retailers in the UK (Tesco and Marks & Spencer) in early 2007. It was implemented in April 2007, when 3 tons of labels were stuck on to packs of imported vegetables. The airplane logo has subsequently been incorporated into existing labels. The idea was to give consumers further information on the product and for them to decide whether to buy air-flown imported produce or not. The concept of 'food miles' was thus implemented. The concept refers to the ecological impacts of food transport, especially long-haul aviation. However, food miles deal with only one aspect of the carbon profile of a product. Both UK retailers admitted seven months later that the controversial stickers had had no impact on sales and that there was no direct evidence to suggest that consumers were so concerned about their footprints that they turned away from air-flown imports (*Fresh Produce Journal (FPJ)*, 2007).

Analysts have since sought to widen the debate by focusing on the carbon footprint of a product. This refers to the 'overall amount of carbon dioxide and other greenhouse gas emissions (e.g. methane etc.) associated with a product along its supply chain and sometimes including use and end of life recovery and disposal' (European Platform on Life Cycle Assessment, European Commission, 2007[15]). This wider definition draws attention to the multiple sources of emissions, which include electricity production in power plants, heating with fossil fuels, transport operations and other industrial and agricultural processes.

The Carbon Trust (2008)[16] has defined a carbon footprint as:

> 'The total set of greenhouse gas emission caused directly and indirectly by an [individual, event, organisation, product] expressed as CO_{2e}'.

After much discussion and debate to come up with a comprehensive method to measure a carbon footprint consistently, the Carbon Trust has developed with the British Standards Institute (BSI) the Carbon Reduction Label, which will inform consumers of the amount of carbon dioxide and other greenhouse gases produced during the full life-cycle analysis (LCA) of the product, including use and disposal. Carbon labelling is a tool designed to provide information to consumers and purchasers who are concerned with the impact of their choices on global warming. It is a mitigation instrument.

Tesco has started the first trial using the Carbon Reduction Label on its own-brand products, and two of the four products chosen are potatoes and orange juice. For the orange juice the most carbon intensive stage is the production part of the crop (*FPJ*, 2008).

In the future, Tesco is planning to restrict airfreight to no more than 1 per cent of its imports, with a bias in favour of sourcing from developing countries. Marks & Spencer has similar plans. Casino, a leading French supermarket, has advised that it will initially focus on the airfreight and packaging aspects of the trade in order to mitigate their climate impact.

The views of other markets and sectors

Other significant retail outlets in the UK and Europe have taken a more practical approach than Tesco. Waitrose will examine operations along the supply chain and introduce measures to reduce the carbon footprint within the UK by transport optimisation and the use of green refrigerants (Bingley, 2008). The Co-operative, in a response to the Soil Association's call for reducing air-flown organics and linking in ethical trade standards, states '... focusing on airfreight is a very poor proxy for the environmental impact of a product, and also does not adequately deal with other social and/or economic consequences of disincentivising airfreight, particularly for producers in the developing world...'. The UK Minister of Trade and Development, Gareth Thomas, expressed similar sentiments when he said: 'UK consumers should not assume that buying food from abroad is worse for the environment than buying from the UK. The livelihoods of a million African farmers are threatened by a misinformed food miles debate' (*FPJ*, 2008).

The Chilled Food Association (CFA) is against carbon footprint labelling of products, which it considers to be misleading and prefers any labelling to apply to businesses and not the products (*FPJ*, 2008).

Comparative carbon efficiency of products around the world

A number of studies undertaken over the last few years have brought out new data on the carbon efficiency of different products. The following section provides some examples.

Fresh fruit and vegetable exports from sub-Saharan Africa (SSA) to the UK account for a maximum 0.1 per cent of total UK emissions (MacGregor and Vorley, 2006). A Cranfield University study found that the carbon footprint of roses produced in Holland was over 5.8 times that of roses produced in Kenya, even after including the emissions from airfreight (Williams, 2007). A simple comparison of outdoor, field-grown green beans in the UK and Kenya showed that the overall CO_{2e} footprint was higher for the Kenyan beans, because of the airfreight emissions. However, field-grown green beans can only be produced in summer in the UK and any year-round production would have to be in heated and lit glasshouses and that would bring about a similar scenario to the roses. According to a report from the New Zealand Agribusiness and Economics Research Unit (AERU), the carbon footprint of seafreighted New Zealand milk solids, lamb and apples sold in the UK is up to four times lower than that of their locally produced equivalent, even with the emissions from transport included (Muller, 2007).

The emissions from airfreight will constitute the highest proportion of a LCA, and a UK study focusing exclusively on the transportation of lettuce, apples and cherries found that UK- and Spanish-grown lettuces had the lowest average CO_2 emissions, whilst the airfreighted cherries that had travelled 7,751kms from North America emitted 80 times more CO_2 than the lettuce. The seafreighted apples that had travelled over 8,000kms emitted 2.4 times more CO_2.

The handling and storage aspect for perishable agricultural produce post harvest also utilises a lot of energy. Research conducted on UK and New Zealand onions shows that the CO_2 emissions associated with the UK storage of locally produced onions are greater than the emissions from seafreighting New Zealand onions to the UK (Saunders, Barber and Taylor, 2006).

Part of a LCA would also have to include the consumer's shopping trip, the cooking of the produce and final disposal of the waste. According to Van Hauwermeiren, Coene, Engelen and Mathijs (2007) a consumer's shopping trip of more than 10kms to solely purchase one kilogram of fresh produce will generate more CO_2 emissions than the airfreighting of the same produce from Kenya.

Box 3.1 The example of Kenya

Developing countries like Kenya are still primarily agricultural-based economies and cannot stop producing and exporting fresh produce. The economic and social ramifications would have a major impact on the economy were this to happen. Kenya is the largest supplier of horticultural produce from SSA and earns over US$750 million in revenue annually. The industry provides jobs to over 150,000 people directly and supports over 4 million people directly and indirectly (EurepGAP, 2007; FPEAK, 2007[17]).

Currently, Kenya supplies one out of every three roses sold within the European Union. On a land mass the size of Texas or France, the country is endowed with a range of altitudes and growing climates that allow for year-round production. The majority of the land has only been farmed in the last 100 years and that too meets with a sustainable approach. The use of fertilisers and agrochemicals is extremely prudent, and with two rain seasons a year, 12 hours of daylight every day and constant temperatures, production requires no heating or light. The majority of the power generated in the country is either hydroelectric or geothermal. The soils are not deep and do not require any intensive machinery nor equipment. Agriculture and farming in Kenya is extremely labour intensive and creates employment for over one million Kenyans in horticulture, tea, coffee and other perennial cash crops. The horticulture industry is also the largest employer of women and of university graduates annually.

Kenya is one of the few countries that are ideally suited for year-round farming activities that have a minimal impact on carbon emissions in the production stages of a LCA.

Issues for developing countries

What effect will labelling have? Since these are private-sector initiatives and not national legal requirements it is difficult to control what the retailers want. Only one third of shoppers are aware of the concept of food miles (MacGregor and Vorley, 2006) and there is little evidence at present that consumers have significantly changed their behaviour and are purchasing more locally produced food (International Trade Center, 2007). A recent report from the United Kingdom Energy Research Centre (UKERC) highlighted that shoppers may be confused by the information contained on the labels. In some cases the consumer may be unaware of the issue highlighted by the label, or overwhelmed by the amount of information provided.

It is important that a full life-cycle analysis of a product, including use and disposal, is undertaken over a series of different production periods and cycles to ensure the information is correctly captured. Identifying and defining the full scope of a LCA is also complex and will need agreement from producers through retailers on each stage. Should certain stages in a LCA receive greater weight for accounting purposes? Once these stages have been defined, the costs of conducting complete LCAs for each stage will need to be calculated. These will be considerable and the total cost implications for a LCA for each product will be significant. Initial funding has been committed by some of the retailers, but the eventual costs will have to be passed on the consumers. In the interim, funding of these LCAs will require resources that a number of developing countries do not have available.

The expense of carrying out LCAs on products will have to be borne initially by producers in conjunction with retailers, with the eventual cost being passed on to the consumers (as stated above). Therefore even though the labelling may seem protectionist and against the spirit of World Trade Organisation (WTO) rules, as long as the customers have a choice to buy between labelled and non-labelled products there is no infringement on fundamental rights. WTO law could be interpreted as not allowing a government to require labels specifying greenhouse gas content in the production process, but there is nothing stopping the private sector from asking for the same. The Technical Barriers to Trade (TBT) Agreement does discourage any methods that would give domestically produced goods an unfair advantage, but it clearly allows non-discriminatory labelling. However, once retailers insist on having only products that have a specific label, there is no product differentiation and this would inhibit free trade and market access would become an issue.

Private standards are taking on quasi-governmental roles, as food safety issues are the prerogative of national governments. When private standards take over these issues, they can interfere with free trade and can become non-tariff barriers, not complying with the spirit of the rules of the WTO. Similarities can be drawn to the private Globalgap standards that retailers have endorsed. Produce that is now not certified under Globalgap does not have access to a number of leading supermarkets, and this effectively blocks producers from market access and free trade and does not give consumers a choice.

According to Appleton, (*Business Law Brief*, 2007) the debate can only be cleared when the definitions of a 'technical regulation' and a 'standard' in Annex 1 of the TBT Agreement are amended. Annex 1(1) and Annex 1(2) both use the phrase 'related production methods' in their first sentence, but fail to use the term 'related' in their second sentences, which govern the 'labelling' of a product, process or production method. This omission leaves room to argue that labelling requirements need not be 'product-related'. The interpretation accepted by most WTO Members is that Annex 1(1) and (2) signify that only 'product-related' process and production methods (PPMs) are covered by the TBT Agreement, and that only labelling requirements that are product-related fall within the Agreement. Carbon emissions from the transport of produce are classified as non-product-related PPMs, as they cannot be detected in the product, and so technically are not covered under the TBT Agreement. Since emissions for produce are primarily in international jurisdiction, WTO Members technically cannot govern nor apply their laws. Therefore, can WTO Members govern the labelling advocated by these private-sector initiatives? At the time of writing not, unless the private-sector initiatives are fully sanctioned by WTO Members. This debate is ongoing in the WTO SPS Committee, but it is highly unlikely that developed country governments would take on the responsibility of private-sector codes.

Reduction in carbon dioxide and other greenhouse gases is of paramount importance. The UNFCCC and the Kyoto Protocol have legally-binding measures in place to ensure that nations reduce their emissions over a set period of time. However, airfreight and sea transportation have been excluded from the reductions required under the protocol (Rigby and Brown, 2003). It is difficult to agree upon how gases emitted by ships should be allocated between countries, as they sail in international waters outside national jurisdictions.

At the same time, the Kyoto Protocol recognises the need for equity and non-restrictive economic development in developing countries to aid their transition to a low-carbon future. The majority of the agricultural producer nations are developing countries. Their national carbon dioxide emissions on a per capita basis are far below the global average of 3.6 tons. The UK per capita emission is 9.2 tons, while that of Africa is 1 ton (MacGregor and Vorley, 2007). As such, few developing countries are subjected to mandatory emission reduction obligations.

It is quite clear that a full life-cycle analysis (including use and disposal, not just the airfreight component) of the product needs to be undertaken over a series of different production periods and cycles to ensure the information is correctly captured before labelling. Airfreight will always contribute significantly to the emissions in a LCA. However, in some cases, emissions attributed to internal transport in the importing countries may be equally high. Emissions from seafreight will play a much smaller role in the LCA.

A complete ban on imports is not practical, and advocating consumption of locally grown food year round is unlikely to help as a large number of products cannot be grown efficiently nor cost effectively without setting up expensive growing environments that would be inefficient and high-energy users. We live in a global economy that trades globally and it is not possible to replace all imports with locally produced substitutes. Farming in developed countries is highly subsidised, and when the subsidies are removed and the farmers correctly taxed for their contribution to carbon emissions, imports would still be competitive. It may be an idea to look at labelling subsidised production in developed countries and letting consumers then decide whether they want to buy heavily subsidised domestic production or opt for a product that has been produced efficiently and with no support.

Should LCA be expanded to include economic and social aspects as well? Should developing countries be involved in defining the carbon footprint, the methodology of measuring it and in the labelling, dissemination of information and education on the issue of carbon footprint? Who will bear the costs of life-cycle analyses in the immediate short term? The answers are not obvious and it is important that clear policy and decisions are arrived at sooner rather than later that capture and convey the correct information in the labelling. It is important to ensure that labelling does not become protectionist or a barrier to free trade and when it does, that the appropriate mechanisms are in place to ensure that the global movement of agricultural produce continues uninterrupted.

Notes

1. Barbados Programme of Action for the Sustainable Development of Small Island Developing States adopted at the Global Conference on the Sustainable Development of Small Island Developing States, Barbados, 1994.
2. Bueno, Herzfeld, Stanton and Ackerman (2008).
3. For examples in the USA see the Biofuels Research and Development Enhancement Act, the Energy Storage Technology Advancement Act of 2007, and the FutureGen – Coal-Fired, Zero-Emissions Electricity Generation Programme.
4. See the United Nations Framework Convention on Climate Change, Articles 3(1) and 4(2)(a), as well as the Lake Victoria Commonwealth Climate Change Action Plan, Commonwealth Heads of Government Meeting 2007 (5).
5. Ibid.
6. See http://www.iata.org/whatwedo/environment/climate_change.htm. [accessed on 14 June 2008].
7. *The New York Times*, 'Can Cruise Ships and Coral Coexist'. Available at: http://dotearth.blogs.nytimes.com/2008/01/15/can-cruise-ships-and-coral-coexist/ [accessed on 19 June 2008].
8. Conservation International, 'New Agreement Works To Balance Cruise Ship Tourism and Conservation'. Available at: http://www.celb.org/xp/CELB/news-events/press_releases/051208.xml [accessed 22 June 2008].
9. See 'How You Can Help: Travel Smart'. Available at: http://www.panda.org/how_you_can_help/at_home/travel/index.cfm [accessed on 14 June 2008].
10. The Caribbean Community Climate Change Centre Presentation delivered at the UNFCCC Pilot Network on Technology Information Centres. See: http://www.ttclear.unfccc.int/ttclear/presentations/Bonn2/CCCCC.ppt [accessed 14 June 2008].
11. See the United Nations Framework Convention on Climate Change, Article 3(5).
12. Economic Partnership Agreement between the CARIFORUM states, of the one part, and the European Community and its member states of the other part, Article 183(3) and (4). Available at: www.crnm.org. [accessed 12 March 2008].
13. Ibid. Article 190.
14. Ibid. Article 183(5).
15. See: http://lct.jrc.ec.europa.eu/eplca [accessed 12 June 2009].
16. UK Carbon Trust (2008) 'Carbon Footprinting'. See: http://www.carbontrust.co.uk/default.ct [accessed 12 June 2009].
17. FPEAK assessment.

4

Technologies

4.1 Standards and Labelling for Energy-Efficient Goods

Paul Waide, IEA, and adapted by ICTSD

There is a growing international consensus that enhanced energy efficiency is the principal means economies have at their disposal to satisfy energy requirements while diminishing unwanted energy-related environmental impacts. Many recent high-level analyses (for example, the World Energy Outlook[1]) have demonstrated there is a very substantial cost-effective potential to improve the energy efficiency of end-use equipment, which if realised would bring large-scale benefits in terms of reduced energy-related greenhouse gas emissions, lower-cost energy services and lower energy dependence. This potential is particularly important for developing countries with high energy-import bills, be they least developed countries or remotely located small island developing states.

However, widely available and cost-effective end-use technologies are still relatively underutilised because of numerous market barriers limiting their deployment. In the absence of direct policy measures aimed at overcoming these barriers, perverse situations can arise when the energy performance of end-use goods is not effectively communicated to final users. Product suppliers therefore have no incentive to utilise technical measures that lower the energy consumption of their products, leading to the utilisation of inefficient products.

Governments are increasingly recognising the need to overcome these barriers through direct policy measures. The most widely deployed policies are mandatory minimum energy-performance standards (MEPS) and energy labelling, both of which have been highly effective instruments to improve equipment energy performance. Comparative energy labelling, where the energy performance of an appliance compared to its peers is visible during its sale, is also being applied.

Labelling helps to complete the market by providing important product performance information that would otherwise be missing, but it does not preclude the sale of inefficient products. MEPS therefore ensure that the worst performing products are excluded from market access. Although such policies were first applied to household appliances, they have also been extended to include commercial equipment, industrial motors, buildings and passenger vehicles. When effectively implemented, such policies have been shown to be highly successful at improving the energy efficiency of equipment.

However, as the large majority of energy-using equipment is traded internationally, these policies have implications for the nature and cost of international trade. The existence of different standards for testing and conformity assessment procedures, as well as national import tariff regimes, add to the cost of trading energy-efficient products that comply with international requirements.

At the same time, testing and conformity assessment procedures are complex and reflect both historic and user difference across nations. Any multilateral decision to develop internationally agreed criteria for relatively energy-efficient goods, based on comparable test procedures, would need to be justified by the expectation that the net benefits of increased trade in goods outweigh the inconvenience of harmonisation. Future energy standards and labelling schemes will need to balance the need for accurate and useful data with the need to be simple, transparent and involve sufficiently low transaction costs to include small developing countries and players.

This section reviews product energy-performance standards and labels as they apply to domestic electrical appliances and then addresses the importance of such standards in addressing climate change. Finally, it considers options to reduce unwanted trade barriers through increased harmonisation and trade liberalisation, with an exploration of the trade-offs involved. The focus is on aspects of interest to least developed countries (LDCs) and small island developing states (SIDS).

Energy efficiency standards and labelling

Public policy in a large number of OECD and non-OECD countries reflects a preference for goods that are more efficient in their use of energy over versions of the same goods that are less energy efficient. Such preferences are manifested through regulations that require a minimum energy performance from household electrical appliances and office equipment, requirements to display labels indicating the product's relative energy performance and voluntary labelling schemes that indicate certain goods as exhibiting superior energy performance to competing products.

In 2006, 57 countries representing 80 per cent of the world's population had energy efficiency standards and labelling programmes in place, with many more countries in the process of developing and expanding these mechanisms. For example, China now has MEPS for 19 product categories, mandatory energy labels for refrigerators and room air conditioners, and voluntary energy labelling for over 19 products. In China alone, these measures are projected to have a potential to save 254 terawatt-hours (TWh) of electricity annually and to avoid 54 gigawatts (GW) of power demand capacity by 2020.

Minimum energy-performance requirements

Today, virtually all OECD countries, as well as several non-OECD countries including China and India, but also Bangladesh, Maldives and Nepal, regulate the minimum energy performance of at least one household or office energy-using appliance. MEPS are the most common approach to remove the least efficient appliances from sale. Other countries (such as the EU, Switzerland, Japan and Korea) have also used both mandatory and voluntary energy-performance targets to attain prescribed energy-performance thresholds as a minimum or average for their products. The nature of the service provided can have a fundamental impact on energy consumption and is reflected in the product categories used to establish MEPS. Many other countries, particularly in South America and Southeast Asia, as well as in parts of non-OECD Europe, Africa and the Middle East, are in the process of developing energy-performance regulations for appliances.

Minimum energy-performance requirements have taken various forms in different countries, with MEPS often coupled with more stringent standards or by more restrictive, voluntary agreements between government and industry. For example, the Korean government sets both MEPS and 'target energy-performance standards' (TEPS), which respectively establish the bottom and top of Korea's mandatory comparative energy labels. In some federal systems of government, such as in the United States and Canada, sub-national MEPS operate alongside federal ones, with local MEPS harmonised for the most part with federal MEPS. The EU has MEPS in place for refrigerators, freezers, boilers and ballasts, while allowing industry to commit to phase-outs of inefficient equipment. The EU has recently enacted a new directive that will give the European Commission the authority to impose MEPS or negotiate voluntary agreements for a very wide range of energy-using equipment without needing to pass primary legislation.

Mandatory energy-information labels

Most countries that regulate MEPS also require energy-information labels to be displayed on the same products. The exceptions, as of 2005, were Chinese Taipei, Ghana and Saudi Arabia, all of which have regulated MEPS but do not yet require energy information labels. Mandatory energy-informational labels are required for all OECD and EU member countries, and by a growing number of (at least 14 at the time of writing) non-OECD countries for at least one product and more often for several.

The main piece of information provided by a mandatory energy-information label is the appliance's estimated energy consumption in kilowatt-hours (kWh) per year, or per operating cycle derived from standard tests. The label typically shows the product group type and size category within which the model should be compared, as well as the energy consumption of the most and least energy-efficient models within the product group for comparison. Increasingly popular is the use of visual aids, such as the EU's colour-coded performance scale from G (lowest efficiency) to A (highest efficiency), to increase quick comparison between different appliances and to identify the most efficient models.

Research into the comprehension and effectiveness of energy labels shows that labels using multiple efficiency categories or classes are more effective at stimulating efficiency gains than those that simply present the efficiency or energy as numerical values or on a continuous scale. Research also shows that there can be significant cultural differences that may render some types of label designs much more effective than others in specific local contexts. This is a powerful reason not to harmonise the appearance of labels without testing that they remain effective instruments at communicating efficiency benefits first. However, it is also important to recognise that the harmonisation of labels allows for cross-country product comparisons. For example, the EU label has been standardised across all EU member states and appears on products sold in Iceland, Lichtenstein, Norway, Switzerland, Russia, Turkey, Bulgaria, Croatia, Romania and South Africa. Other countries, including Brazil and China, have utilised some aspect of the EU energy label in their own labelling schemes, but despite this voluntary adoption across many countries, there are others that have adopted different labelling formats.

'Seal-of-approval' and other voluntary labels

'Seal-of-approval' labels are voluntary and selective, and are awarded only to products that meet relatively strict environmental requirements. Many labels are administered by governments and closely co-ordinated with mandatory energy labelling programmes, for example, the EU's Eco-label award scheme, China's Great Wall energy certification label and the US's Energy Star programme. Others are administered by non-profit organisations, for example, Japan's Eco Mark scheme and the US's Green Seal, or by representatives of governmental and non-governmental organisations.

Where a mandatory MEPS already exists for a given product, the minimum energy-performance level that the same product must meet is set at a value representing anywhere from a 10 per cent to a 55 per cent improvement over its corresponding MEPS value. Ecological or environmental labelling schemes ('eco-labels') often augment energy-efficiency criteria with other product performance criteria, such as for noise, water use or product durability and specifications relating to the composition of the product or the product's re-use, recycling and disposal characteristics. The criteria set by the EU Eco-label for portable computers, for example, include restrictions on the use of harmful substances, such as flame retardants, heavy metals and plastics; criteria for durability; and end-of-life criteria. Most private third-party voluntary labelling schemes contain similar criteria unrelated to the product's energy performance.

Determining product energy-performance

Energy performance is not a universally defined quotient, and each class of appliance requires its own method of measurement. While some energy-efficient goods can be distinguished through simple physical inspection, this is usually insufficient because relative energy performance, with the exception of groups such as broad lamps and liquid crystal display (LCD) computer monitors, is not apparent from appearance. The energy-test procedure provides an alternative, standardised method for measuring the energy performance of an appliance and piece of equipment. Well-designed test procedures should be inexpensive, accurate and closely reflect operating conditions, though compromises are made in reality.

However, across countries, these methods and their associated performance requirements can often vary in ways that are not always trivial. Such differences are recognised in the UN Framework Convention on Climate Change, which notes in its preamble that 'environmental standards ... should reflect the environmental and developmental context to which they apply'. These difference pertain to:

- *How countries classify and describe products to regulate MEPS or energy labels.* Because of cultural, commercial and historical factors, the features available on and the configurations of basic household appliances, particularly refrigerator-freezers and washing machines, may exhibit wide regional differences. Variability stems from individual standards or labels developed for categories within a product group.
- *The test procedures used to measure energy performance.* International standards exist for most household appliances and office equipment types, but national differences complicate the development of test procedures that allow both for local flexibility and for effective cross-country comparison. However, interest in making test procedures better reflect local conditions led many countries to adapt standards of the International Organization for Standardization (ISO) and the International Electrotechnical Commission (IEC).
- *The ways to specify standards.* Even when countries apply identical test procedures to measure energy consumption, they do not necessarily apply identical energy efficiency metrics. For example, formulas for calculating an appliance's energy performance may differ in how they adjust for functionality variables such as storage volume or cleaning performance. Differences complicate product comparability.
- *The stringency of required product energy-performance thresholds.* Even where formulas for specifying efficiency requirements are similar, differences often exist in the stringency of efficiency thresholds specified for MEPS, labels or targets. This may reflect diversity in the efficiency of product markets, varying levels of policy ambition, variations in energy prices or simple differences in the timing of introducing regulations.
- *The scheduling of reviews of regulations and test procedures.* Innovation for household and office electrical equipment is often faster than for household goods because of the incorporation of digital technology. Energy-efficiency standards must keep pace with technology changes that affect design and performance in order to remain relevant. While government regulations often mandate reviews of MEPS and labels every three to five years, different starting dates results in review cycles out of sequence.

The reconciliation of these differences would be a major undertaking. While some of them arise from the development of independent regulatory traditions, others reflect genuine local physical differences. For example, the energy consumption of a refrigerator or air conditioner is strongly dependent on local ambient temperatures and is therefore sometimes reflected by differences in energy performance test procedure rating conditions.

Any multilateral decision to begin developing internationally agreed criteria for relatively energy-efficient goods, based on comparable test procedures, would therefore need to be justified by the expectation that the net benefits of increased trade in goods outweigh the inconvenience of harmonisation. It is important to identify the degree of existing commonality, the sources and rationale for identified differences, and the institutional and financial cost of producing a common international performance rating system. Analysis would depend on the potential size of each product's international market, the product's contribution to world energy consumption and the degree to which tariff and non-tariff barriers are currently restricting trade.

The benefits for climate change

After transportation, household and office electrical appliances represent the world's fastest-growing segment of total energy consumption. World purchases of major appliances and equipment increased by roughly 3.7 per cent a year in the decade from 1992 to 2002, and are continuing to grow. While much hope has been placed on low- and zero-carbon alternatives such as hydropower, nuclear power and renewables, these methods face cost, availability or other constraints that limit their rate of adoption. In contrast, improved end-use energy efficiency offers the principal opportunity to lower the rate of CO_2 emissions in the near term without significant increases in energy services cost. In fact, alternative energy development is credited only with 12 per cent of emissions reduction savings potential by 2030, compared to the 70 per cent

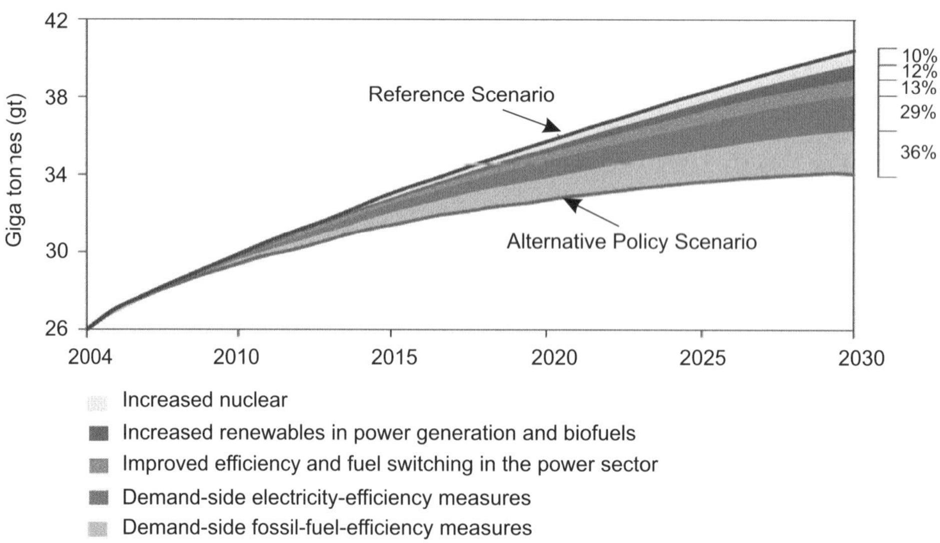

Figure 4.1 Global savings in CO_2 emissions in alternative policy scenarios compared to the reference scenario

Source: IEA, World Energy Outlook (2006).

reduction of accelerated energy efficiency improvement according to scenarios developed by the IEA (2006b).

Current standards and labelling schemes have been highly effective at reducing energy demand and carbon emissions. The International Energy Agency (IEA) has estimated that appliance efficiency policies put in place in OECD countries between 1990 and 2002 were on course to save 292TWh of residential electricity demand in 2010 and 393TWh in 2020 (some 13.5 per cent of the forecast total residential electricity consumption). This is set to avoid over 146 million tons (Mt) of CO_2 emissions per year in 2010 and 240Mt CO_2 per year in 2030 at a net benefit to society of between US$69 and €169 per tonne of CO_2 avoided depending on the region. In the case of lighting, the impacts of energy efficiency standards and labelling schemes are even more impressive. The IEA estimates that globally current policies saved 334TWh of electricity demand in 2005 and are on course to avoid 745TWh of demand in 2030. The same policies are projected to have saved 1.6 billion tonnes of cumulative CO_2 emissions from 1990 to 2005 and to be on course to save a further 8.6 giga tonnes (Gt) to 2030 even without further strengthening beyond current settings.

The imbalance in current supply-side and demand-side incentives is such that there is a large share of current energy demand and energy-related carbon emissions that could be avoided without any loss of energy service and at net negative cost to society. There is no reliable quantification of how large this opportunity is across all end-uses in the global economy, but good estimates exist for specific sectors. The power sector is the main source of energy-related CO_2 emissions and this power is used in a variety of electrical end-uses. The cost-effective savings potentials in these end-uses vary from almost 40 per cent for lighting, which accounts for about 19 per cent of global electricity use (IEA, 2006a), to about 25 per cent for motors and drives, which account for about 40 per cent of global electricity use. In the case of household appliances in the OECD, it has been estimated that 35 per cent of domestic appliance electricity use could be saved cost-effectively (IEA, 2003). Similarly impressive cases have been documented in other end-use sectors in non-OECD markets alike.

The World Energy Outlook Reference Scenario (IEA, 2006b) suggests that over US$10 trillion will be needed for the electricity sector in 2030. Avoiding 33 per cent of future electricity demand would decrease the need for US$3.3 trillion in electricity sector supply-side investments. It would cost between 3 and 6 times as much to purchase a kWh of electricity than it would cost to avoid needing it at all through preferential purchase of more efficient electrical equipment. Overall, it is probable that about 30–35 per cent of global electricity demand could be avoided through the preferential use of currently available efficient end-use technologies.

The level of saving can be significant in many developing countries, which have rapidly growing populations and increasing energy needs. In many island and remote countries that depend on imports of energy, demand-side measures can contribute a significant portion of energy savings and reduce the costs associated to energy imports.

With this evidence, standards and labelling schemes therefore hold tremendous potential in reducing energy use and related carbon emissions.

Implications for international trade

The immense potential of energy-efficient products to reduce world carbon emissions has led some to call for the increased promulgation of these goods. There is still considerable scope for facilitating trade and reducing emissions by lowering both tariff and non-tariff barriers to trade. Different efficiency standards and labels, as well as test procedures, are reported to be the most frequent concern of the World Trade Organization (WTO) under the Agreement on Technical Barriers to Trade. Measures available to parties in the implementation of the Kyoto Protocol, such as energy standards and labels, may have trade effects. WTO rules, through disciplines on areas such as subsidies, border measures and technical requirements, therefore have crucial implications on the options available to countries in implementing climate measures and promoting the use of energy-efficient products.

Efforts to promote efficient and cost-effective products therefore need to consider the consequences and trade implications of trade liberalisation for these goods, with possible solutions to be found in the harmonisation of standards and in the reduction of high tariffs on energy-efficient products, while recognising the need for flexibility to adapt to local circumstances. While there are certainly costs in restructuring testing, standards and labelling schemes, it can be argued that the benefits of the resulting energy savings are much greater and thus provide compelling motivation for lowering barriers to trade.

Harmonisation of standards

Since for the majority of cases, energy-efficient goods cannot be distinguished simply by their appearance, enforcement of regulations requires that performance can be determined through testing. The most immediately obvious approach would be to require a common efficiency threshold to be attained, regardless of where the good is sold. This would require: agreement on the use of either common or mutually convertible energy-performance test procedures; either common, or mutually convertible energy-efficiency metrics and product categories; and common energy-efficiency thresholds. This would also require taking into account differences in technical capacities, especially in smaller developing countries, to align their test procedures and performance thresholds with international standards.

For energy-efficient goods to be included in tariff-reduction or tariff-elimination initiatives, there would have to be agreement on a standard set of product descriptions, reference testing standards and efficiency thresholds etc. This implies a major effort toward harmonisation, which would generate greater coherence at the international level. Whether that can be achieved under the UNFCCC umbrella or through other fora such as the ISO requires further consideration. Overall, views diverge on the desirability and need for harmonisation of energy standards at the international level.

Some argue that this is a sphere for 'regulatory competition' where consumers play an arbitration function. A key challenge for smaller developing countries is to ensure that 'regulatory competition' does not result in hindering market entry for products that may be less performing, and that effective flexibilities and transition periods are provided for such countries as may be required.

International experience suggests that there already is a convergence of test procedures in several product groups. Among countries having energy labels or MEPS for computers, almost all use the US Environmental Protection Agency's Energy Star test procedure, with the exceptions using test procedures very similar to those of the EPA. Similarly, the vast majority of countries use the same international test procedures for compact fluorescent lamps (CFLs), namely International Electrotechnical Commission (IEC) 60901 and IEC 60969. However, performance tests used for voluntary CFL energy-performance labelling schemes are quite diverse, with over 30 variations in requirements being identified through the international CFL harmonisation initiative. There also remain variances in test procedures, such as for refrigerator-freezers, where different regional and national test procedures in use specify incomparable and different means of measuring internal temperatures. Appliances are designed with the satisfaction of specific test procedures in mind, meaning that their performance is usually optimised to that test procedure. For example, a refrigerator-freezer that is optimised to perform well under the Japanese test procedure, which includes door-openings during the test, might place relatively greater design emphasis on an efficient refrigeration system compared with one that is optimised for other test procedures where the door is not opened and the quality of insulation plays a relatively greater role.

These trends are also similar across products for product categories and efficiency metrics, with a common basic methodology used to define energy efficiency across countries. However, the significant differences in definition in product categories such as for refrigerator-freezers, which in turn reflect the different demands of energy test procedures in different countries, results in a basic problem of comparability even once differences in units are adjusted for. Furthermore, the energy consumption of a specific appliance type is often measured against an equivalently featured appliance with average energy performance at some point in the past. The differences in the evolution of national markets means that these are not easily comparable, although in recent years there has been a tendency for countries developing new metrics to harmonise them with existing ones. Efforts toward common reference energy-performance standards hold great promise in lowering the non-tariff, technical barrier to trade. However, in aligning these processes, policy-makers should take care not to threaten legitimate functional distinctions and the integrity of standards and labelling schemes.

Tariff concerns

Facilitating the use of energy-efficient goods also requires examination of the tariff barriers to their trade. A joint study by the IEA and OECD[2] on energy-efficient electrical appliances found that the market is large, with demand for household appliances likely to grow rapidly. However, tariffs on many household appliances remain high, at least in some developing countries – up to 50 per cent in some African countries, but 0–10 per cent in most OECD countries. In general, LDCs tend to have the highest levels of tariffs on energy-efficient electric appliances, making the removal of such tariffs an economically and politically sensitive matter. The joint study also noted that the scale of environmental outcomes from liberalisation is less easy to determine.

In the December 2007 United Nations Climate Change Conference in Bali, the US and EU submitted a joint proposal for trade liberalisation in climate-friendly goods and services. The elimination of trade barriers facing energy-efficient goods and services could lower costs and promote increased use. However, proposals for trade liberalisation of climate-related goods continue to receive mixed responses, with developing countries expressing concern about reducing trade barriers (see section 5.2, below).

While tariff reduction on energy-efficient goods remains a hotly debated issue on the agenda, proposals for process and production method- (PPM-) based tariffs are also being discussed. As mentioned before, the production process of a product contributes to its overall GHG emissions and therefore has raised calls to consider PPM when evaluating efficiency. Including tariffs on goods produced in inefficient ways would restrict the promulgation of these goods and further promote the use of energy-efficient products. However, there is considerable resistance to PPM-based measures in the WTO, especially voiced by developing countries. PPM-based measures raise the question of equity, as they are most frequently used by industrialised, importing countries. Meanwhile, such tariffs and measures would place increased burden on producers from developing countries, as their products are denied entrance because of PPM considerations. Given the difficulty of financing innovation and the barriers to trade, PPM-based measures would make it more difficult for developing countries and LDCs to export their goods and therefore effectively build the capacity to produce more energy-efficient goods.

Table 4.1 Tariffs on energy-efficient end-use appliances

	LDCs (Low income)	Middle income	Transition Economies	High income non-OECD	OECD and non-OECD EU
Refrigerator – freezers	24.5%	22.3%	12.2%	3.8%	2.1%
Room Air Cons	24.9%	17.5%	10.4%	2.5%	1.9%
Fluorescents	26.2%	14.4%	11.3%	2.4%	3.4%
Incandescents	17.7%	15.6%	14.7%	1.0%	3.8%
PCs	9.7%	1.3%	3.5%	0.1%	0.3%
Laptops	9.6%	1.2%	1.6%	0.1%	0.0%

Source: Steenblik, Vaughan and Waide, 2006.

There are also several legal questions regarding the use of PPM-based measures to promote energy-efficient products. Most importantly, the obligation not to discriminate between 'like' products under the GATT raises the question of whether WTO Members can favour one product over another based on the GHG emissions of its production. Second, Members question whether energy-efficient products fall under the GATT's general exceptions clause, which can justify environmental and health measures. The debate surrounding PPM-based measures is still less than concrete, but the Appellate Body has indicated that measures must be flexible to allow for compliance, while remaining transparent and fair.

In conclusion, incentivising the development of more energy-efficient technology and promulgating its use can provide a powerful solution for reducing emissions. There remain many issues to consider when promoting the use of energy-efficient products, but policy-makers must keep in mind the challenges and implications, especially for developing countries, of creating immediate reforms oriented toward more technologically innovative products.

4.2 Liberalisation of Environmental Goods and Services

Mahesh Sugathan, ICTSD

Trade is an important channel for the diffusion of many climate mitigation technologies and goods. Few countries have the domestic capacities or know-how to produce all that they need. This is particularly true for many developing countries, including least developed countries, small island developing states and small and vulnerable economics (LDCs, SIDS and SVEs). Although building domestic capacities may be their long-term goal, trade liberalisation can provide rapid access to key technologies. Trade liberalisation – whether locked in through negotiations at the World Trade Organization (WTO) or elsewhere, or undertaken autonomously – can also lower the costs of environmental goods by allowing consumers (industries or households) to purchase them at world market prices.

A 2007 World Bank study, *International Trade and Climate Change*, points to the potential for liberalisation in the area of low-carbon goods to lead to real increases in trade flows. According to World Bank estimates, the removal of tariffs for four basic clean-energy technologies (wind, solar, clean coal and efficient lighting) in 18 developing countries with high greenhouse gas emissions would result in trade gains of up to 7 per cent. The removal of both tariffs and non-tariff barriers could boost trade by as much as 13 per cent. The net effect would, however, vary across technologies and across countries, depending on existing barriers and the import elasticities of demand.

Paragraph 31(iii) of the Doha mandate, agreed by all WTO Members in 2001, calls for a reduction or, as appropriate, elimination of tariffs and non-tariff barriers on environmental goods and services. This mandate offers a good opportunity to put climate-friendly goods and services on a fast track to liberalisation, although, as the negotiations to date have shown, this is not a simple proposition.

Many developing countries are, however, reluctant to make such blanket commitments. They are concerned that they might, unintentionally, end up liberalising far more goods than just those with an environmental end use. They are also concerned about competition for their own small and medium-sized enterprises (SMEs) or possible future companies producing the same goods. Furthermore, as developing and LDC countries have in general higher tariffs on environmental and energy-efficient goods than Organisation of Economic Co-operation and Development (OECD) countries, tariff cuts have large implications not only on the market potential in developing countries, but also on government revenue from import taxes.

With regard to least developed countries, it is clear that possibilities of including products of export interest to them may largely be confined to environmentally preferable products (EPPs). However, few products, if any, may be explicitly linked to climate change mitigation objectives, except indirectly – for instance, EPPs harvested or gathered sustainably from rainforests.

Overall, trade liberalisation by itself may not be sufficient to promote the diffusion of climate-friendly goods in smaller developing countries. A whole host of complementary measures – regulatory, capacity building, financial and technology-related – will be required.

This section will survey the key issues surrounding liberalised trade in low-carbon goods from the point of view of developing countries. The focus will be on climate mitigation goods rather than those relevant to adaptation. For many developing countries, adaptation is a higher priority than undertaking mitigation measures. However, adaptation goods and technologies are diverse and diffuse and often involve low-tech local solutions and materials. For these reasons, they have not yet been a priority in discussions on environmental goods and services (EGS) liberalisation.

Issues in environmental goods and services negotiations and challenges for developing countries

The Doha Round negotiations have met with challenges in a number of areas – many of key importance to developing countries – as highlighted below.

Defining and classifying climate-friendly goods

The lack of a universally accepted definition of environmental goods has slowed down agreement on product coverage in negotiations on environmental goods. Two broad categories of environmental goods have featured in the WTO discussions so far: traditional environmental goods, with the main purpose of addressing or remedying an environmental problem (e.g. carbon capture and storage technologies); and environmentally preferable products (EPPs), which include any product with certain environmental benefits arising either during the production, use or disposal stage relative to a substitute or 'like' product. Figure 4.2 below provides some examples of products from both categories.

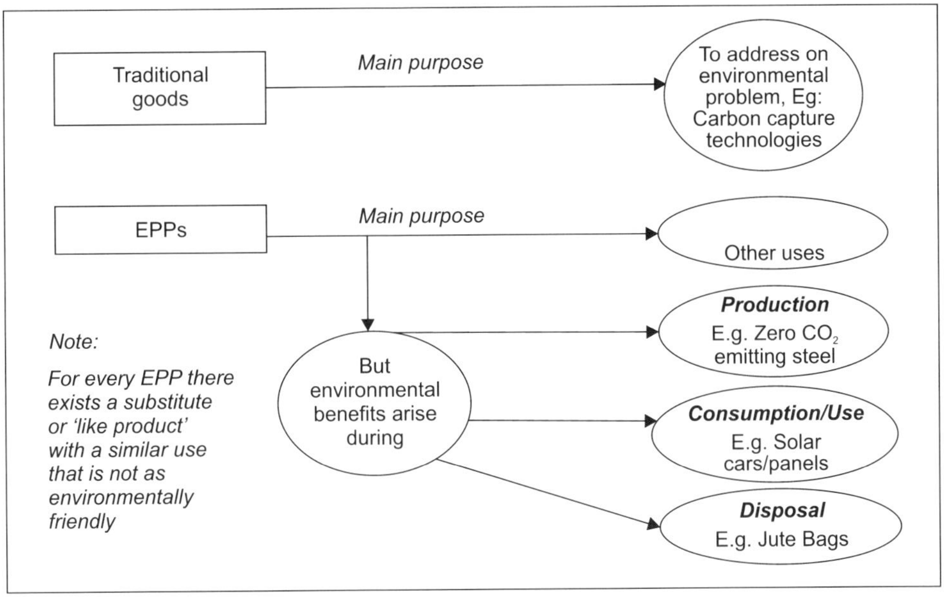

Figure 4.2 Traditional goods vs. environmentally preferable products (EPPs)

Source: Claro et al., 2007.

Introducing an additional layer of complexity, products can be environmentally preferable, either due to improvements in embedded technology (e.g. more energy-efficient variants of the same good, such as a car) or as compared to a different product (such as solar cookers versus wood-burning stoves).

In terms of classification, categories and sub-categories of goods are assigned a code within the Harmonized Commodity Description and Coding System (HS), allowing countries to track trade volumes and tariff levels. The more digits are included in a code, the more specific the description of the good. At the WTO, countries have HS numbers for products only up to the six-digit level. Beyond that, as product descriptions get more specific, different Members use different codes and descriptions. This makes it difficult to clearly identify environmental goods, including climate mitigation goods, at the six-digit level. They are often lumped together with other goods that are unrelated to the environment or climate mitigation. For example, one list of proposed products contains HS-8413.81: 'pumps for liquids, whether or not fitted with a measuring device; other pumps'. Such pumps are often used by wind turbines for energy storage, but at the six-digit level of generality it is impossible to separate those pumps used in this manner from pumps used in any number of other applications. While it is possible to identify and liberalise specific goods using 'ex-outs' beyond the HS-6 digit level, member countries need to agree on product codes, or at least product descriptions in the area of climate mitigation, which can be a time-consuming process.

'Process and production methods (PPMs),' relativity and evolving technology

Most WTO Members have not accorded 'environmental goods' status to otherwise 'like' products that have been produced using methods friendlier to the environment. This is due to the difficulty of distinguishing such products within the HS system and challenges of harmonising standards and labelling, as well as to systemic concerns with regard to other non-product-related standards making their way into the WTO system as a basis for differentiated treatment. Even for products where the environmental benefits do not depend on PPMs, many are only relatively eco-friendly. Hybrid cars, which can be compared to electric cars, provide one example. Moreover, technological change could make existing 'relatively friendly' environmental goods obsolete tomorrow. How should trade negotiations respond to these challenges? Once lowered and bound, tariffs cannot be raised again for obsolete products. At the very least, newer products that emerge should automatically benefit from the trade benefits accorded to the obsolete one. If relatively clean goods are accorded preferences, should we distinguish based on national-level baselines, or some internationally set baseline? Predominant methods of production differ dramatically across countries. Some experts, including Mytelka (2007), argue that only truly 'clean' technologies should benefit from EGS liberalisation – as opposed to 'relatively cleaner' products, but then we are left with the challenge of defining truly clean – particularly challenging as one takes a longer-term perspective.

The dual-use problem

The dual-use problem is one of most important challenges facing environmental goods negotiators. It arises from the fact that most product categories proposed by WTO Members as environmental goods for rapid liberalisation include, at the HS-6 digit level, other products that also have non-environmental uses. In other cases, a specific ex-out product, such as a pipe, may intrinsically be dual-use and used for environmental and non-environmental purposes. Pipes, for instance, are used as components of sewage treatment plants as well as for transporting oil and gas. The two types of dual-use products are illustrated below.

Most developing countries are hesitant to liberalise bound tariffs on dual-use products such as valves and pumps due to concerns about the impact of such overarching liberalisation on their established domestic industries. Proponents of these liberalisation efforts argue that the environmental benefits would be limited if liberalisation was confined only to a handful of products used solely for environmental purposes. For a number of developing and least developed countries in particular, the tariff revenue gains from many of these products may be important and this needs to be taken into account.

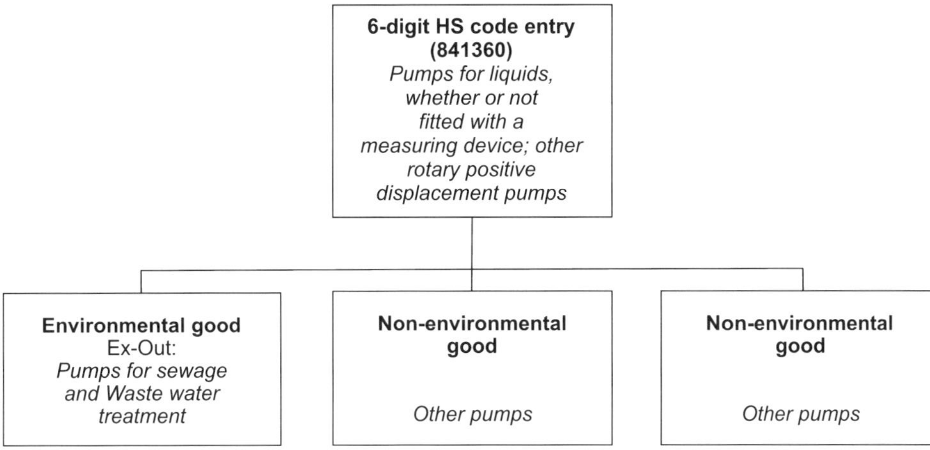

Dual-Use: Type 1

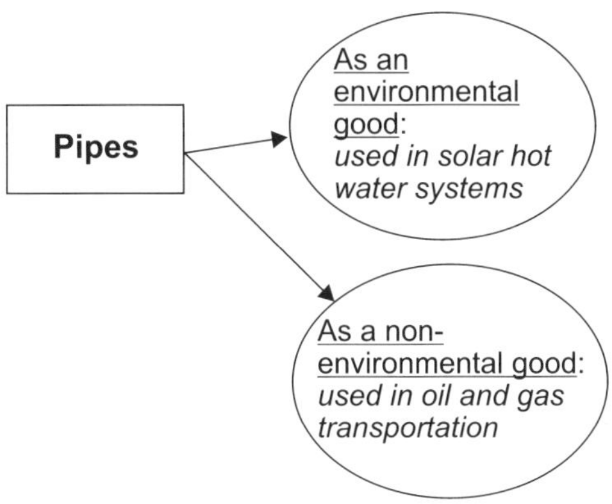

Dual-Use: Type 2

Figure 4.3 Types of dual-use products

The distribution question

A big challenge for the environmental goods negotiations is to include products of export interest to developing countries. The perception so far has been that environmental goods – being capital- and technology-intensive – are of export interest only to developed countries and a few middle-income developing economies. Others, such as Hamwey (2005), see significant export opportunities for developing countries in a large number of lower-tech environmental goods, such as parts and components. However, these also happen to be the 'dual-use' products with which most developing countries have concerns.

Undoubtedly, many developing countries such as China and India have emerged as leading producers in clean energy sectors such as wind and solar energy, while Brazil is a world leader in biofuel manufacturing equipment. According to the World Bank (2007a), exports of clean-energy products such as efficient lighting are growing rapidly from many developing countries. Analysis by Jha (2008) reveals that China and Mexico were among the top 10 exporters in various categories of environmental goods relevant to climate change mitigation discussed in the WTO. On the other hand, interest in the inclusion of agricultural products early on in the Doha negotiations from Kenya, Ghana, Mauritius and other African countries and later on by Latin American countries, particularly ethanol by Brazil, has met with some degree of resistance by traditional developed-country environmental goods proponents. With regard to least developed countries, it is clear that possibilities of including products of export interest to them may largely be confined to environmentally preferable products (EPPs). However, few products, if any, may be explicitly linked to climate change mitigation objectives, except indirectly – for instance, EPPs harvested or gathered sustainably from rainforests. In any case, these may bring up PPMs again or such products may not be significant in terms of generating export revenue.

Approaches to liberalisation

In addition to issues of product coverage, the question of how to approach the liberalisation exercise has been another big stumbling block to progress in the Doha Round negotiations on environmental goods and services. For many developing countries, this issue needs to be resolved before the talks can progress to product coverage. Fundamentally, many developing countries are unwilling to commit to bound liberalisation on lists that comprise mostly dual-use products. Some have therefore proposed their own alternative approaches to liberalisation.

The *List Approach* is favoured by the so called 'Friends of Environmental Goods', which comprises Canada, the European Union, Japan, Korea, New Zealand, Norway, Chinese Taipei, Switzerland and the United States. The approach essentially consists of identifying and submitting lists of what Members regard as environmental goods of interest for accelerated and permanent liberalisation by reducing or eliminating bound tariffs. India's *Project Approach*, meanwhile, proposes liberalising any good or service intended for a specific environmental project as approved by a Designated National Authority for clean development mechanism (CDM) project activities and based on criteria developed by the WTO's Committee on Trade and Environment. Such liberalisation would be temporary, lasting for the duration of the project, and domestic implementation of the criteria would be subject to WTO Dispute Settlement. The *Integrated Approach*, proposed by Argentina, resembles the project approach, but with further identification of goods used in the various approved projects. Both approaches are driven by concerns of ensuring 'environmental end-use' of products that are mainly dual-use. A fourth approach – the *Request Offer Approach* – has been proposed by Brazil, whereby countries would request specific liberalisation commitments from each other on products of interest to them and extend tariff cuts they deem appropriate equally to all WTO Members. Some Members have informally proposed combining various approaches, depending on whether the good in question is single or dual-use. At the time of writing, there appears to be no resolution on which approach or combination of approaches to follow.

The World Bank report (2007a) has proposed accelerated liberalisation of products, technologies and services used in CDM projects. According to the report, such liberalisation could reduce equipment costs and contribute to lowering transaction costs for potential investors, as long as they were complemented by certain measures, such as supportive local regulatory measures.

Technology transfer and special treatment of developing countries

During the course of negotiations, many countries, including China, have stressed the need to facilitate technology transfer. Canada, among others, has stressed technology transfer as occurring through aid, private investment, technical assistance, partnerships between research organisations and small companies, and trade in environmental technologies themselves. Others, such as Cuba, prefer a differentiated treatment for developing countries, including transfer of technologies on favourable and preferential terms with related know-how and necessary training. Lack of adequate attention to technology transfer remains one of the main complaints with regard to the list approach. No WTO member has, however, proposed a practical way to operationalise technology transfer through WTO environmental goods and services negotiations.

Other cross-cutting issues that have been raised during environment goods discussions include the need to identify and deal with non-tariff measures and ensure special and differential treatment (S&DT) for developing countries. Various S&DT proposals – such as multiple product lists with different rates of tariff reduction, sensitive product exemptions and longer implementation periods – have been made by various WTO Members.

Climate-relevant proposals

From a climate mitigation perspective, the environmental goods negotiations have seen proposals from Qatar, the Friends and more recently from the United States and EU, which have included 'climate-friendly' goods. Early on in the Doha negotiations, Qatar proposed liberalising natural gas-fired generation systems and advanced gas generation systems, citing a reference to its benefits under the UNFCCC. Qatar also referred to the Intergovernmental Panel on Climate Change (IPCC) Assessment Reports, which recommended increased use of natural gas over other fossil fuels as a way to reduce greenhouse gas emissions.

The Friends proposed a list of 153 products, which included categories such as renewable energy products, solid waste management, and heat and energy management products. On 30 November 2007, the United States and EU proposed accelerated liberalisation of goods and services relevant to climate change mitigation, including zero tariffs by 2013 for 43 products that were identified by the World Bank from the Friends 153 list as being relevant to climate change mitigation. There were to be longer phase-in periods for liberalisation by developing countries and participation was made optional for least developed countries. The list of 43 goods included a wide variety of products such as solar collectors and system controllers, wind-turbine parts and components, stoves, grates and cookers and hydrogen fuel cells. The list was supposed to be a starting point for discussions rather than an exhaustive one. The United States and EU further suggested the negotiation of an innovative Environmental Goods and Services Agreement modelled on the existing WTO Information Technology Agreement (ITA) that would include other, non-climate related environmental goods as well. Relevant climate mitigation services such as engineering, maintenance and technical testing were also covered.

Despite the United States pointing out that it was a net importer of these 43 goods and that developing countries such as China, Mexico, Malaysia, Chinese Taipei and Indonesia were among the top exporters, many developing countries questioned the 'development dimension' of the proposed list. Brazil criticised the exclusion of ethanol from the list, and many other developing countries were concerned that the 'climate goods' list, as with most other environmental goods proposed in the WTO, included dual-use products.

Addressing other drivers of trade in poorer developing countries

Overall, it is important to recognise that trade liberalisation by itself may not be sufficient or may only have a miniscule impact on climate mitigation. A whole host of complementary measures – regulatory, capacity building, financial and technology-related – will be required. In this regard, analysis of the Friends' 153 environmental goods list by Jha (2008) is revealing. Jha clearly shows that demand for these products may be determined by factors other than tariffs, such as gross domestic product (GDP), foreign direct investment, enforcement of environmental regulations (shown by environmental performance indices) and the number of bilaterally funded 'environmental' projects. For instance, many African countries already have very low tariffs on many environmental goods, but little or no imports because their GDPs are constrained and they have other import priorities. Trade liberalisation with a lack of purchasing power will certainly not help. In addition, according to Jha (2008), technical assistance or tied-aid projects also appear to be directed towards those countries that have the relevant purchasing power. This gap in environmental goods imports in a large number of developing countries also points to the need for technical assistance projects in developing countries, especially in Africa. Bilateral and multilateral donor assistance in this regard have focused especially on relatively high-income developing countries, notably China, the Republic of Korea, Brazil and Mexico.

Further, while categories within the 153 list that are relevant to climate change mitigation, such as renewable energy and heat and energy management products, appear sensitive to tariffs, long-term dynamic comparative advantage in these products lies (until 2015) with developed countries (for renewable energy) and with middle-income developing countries (for heat and energy management products). It is thus important to ensure that benefits from trade liberalisation also accrue to the poorer developing countries that may either lack resources to import such products or the capacity to produce, operate and deploy them. For renewable energy as well as heat and energy management (HEM) products, no Commonwealth country except for Malaysia figured among the top 10 global exporters or importers in 2006. However, considering developing countries exclusively in 2006, Malaysia ranked fifth among top-ten developing country exporters in the renewable energy category and fifth among top-ten developing country exporters in the HEM category. South Africa was the seventh largest developing country exporter in the HEM category. In terms of 2006 imports, Malaysia was the eighth and India the tenth largest developing country importer in the renewable energy category, while Malaysia was fifth, South Africa seventh and Pakistan the ninth largest developing country importer in the HEM category.

In the category of solid waste management, which many experts regard as being relevant to climate change mitigation (due to greenhouse gas emissions from landfills), Malaysia was the fourth and South Africa the seventh largest exporter among developing countries, while among importers Malaysia, South Africa and Pakistan ranked fourth, seventh and eighth respectively.

For those smaller Commonwealth developing countries that are among the top 10 importers of all environmental goods in the '153' list, namely Malaysia, South Africa and Pakistan, the bound tariff rates in 2006 were 10 per cent (for Malaysia and South Africa) and were unbound for Pakistan. However, the applied tariffs were lower at 4.7 per cent for Malaysia, 2.3 per cent for South Africa and 10 per cent for Pakistan.

Intellectual property rights may also act as a barrier to access, particularly in emerging climate technologies. Trade liberalisation alone may not result in 'take-off' of a technology in developing countries, if costs are kept high due to high licensing fees or royalty payments.

From a long-term perspective, it will also be essential to help developing countries build up their productive and technological capacities in this area. The World Bank report calls for smarter trade as an adjunct to freer trade, and proposes bundling trade liberalisation with a package of technical and financial assistance. The impact of trade liberalisation for climate change mitigation efforts, as with most other sustainable development objectives, will be only be as effective as the broader enabling framework within which it is put into play.

4.3 New Technologies and Innovation

Maria Julia Oliva, ICTSD

Development and transfer of technology has emerged as a basic building block in the crafting of a post-2012 global regime on climate change. A range of technologies needed for mitigation and adaptation to climate change have already been identified. These include technologies needed for observation and monitoring of climate change, technologies for mitigation (e.g. energy-efficient and renewable-energy technologies, energy efficiency transportation technology, energy- and material-saving building and construction technologies, low-greenhouse gas (GHG) emission technologies for agriculture and animal husbandry etc.) and technologies for adaptation (e.g. water-saving, water-capture and water-reuse technologies, agricultural biotechnology, disease and pest-control technology, flood, drought, sea-level rise, agricultural disasters and desertification-control technologies).

Several barriers to access to these technologies have been identified, including economic, institutional, policy and human capacity-related barriers. Smaller developing countries are confronted with many such barriers to the development and transfer of technology. Legal and policy measures have an important role in the transfer of technology, even as technology is largely transferred by the private sector. Trade liberalisation, per se, is an insufficient driver to the diffusion of the knowledge and technologies that will be required, especially in developing countries, to mitigate and adapt to climate change. A range of economic and trade-related instruments provide opportunities for multilateral action to promote climate-relevant innovation and technology transfer, providing an 'enabling environment'.

Technological solutions are one of the imperative elements in meeting the challenges of climate change. A critical factor in GHG emissions, technology is also fundamental to enhancing existing abilities and lowering the costs of reducing these emissions. In this context, the UN Framework Convention on Climate Change (UNFCCC) and the Kyoto Protocol already require Parties to promote and co-operate in the development and diffusion, including transfer, of technologies that control, reduce or prevent GHG emissions[3]. Enhanced action on technology development and transfer will also be central in enabling the full, effective and sustained implementation of the UNFCCC beyond 2012, as recognised in the Bali Action Plan (see box 4.1).

Box 4.1 Technology transfer in the Bali Action Plan

The Bali Action Plan launched 'a comprehensive process to enable the full, effective and sustained implementation of the Convention through long-term co-operative action', by addressing, inter alia:

...'(d) Enhanced action on technology development and transfer to support action on mitigation and adaptation, including, inter alia, consideration of:

(i) Effective mechanisms and enhanced means for the removal of obstacles to, and provision of financial and other incentives for, scaling up of the development and transfer of technology to developing country Parties in order to promote access to affordable environmentally sound technologies;

(ii) Ways to accelerate deployment, diffusion and transfer of affordable environmentally sound technologies;

(iii) Co-operation on research and development of current, new and innovative technology, including win-win solutions;

(iv) The effectiveness of mechanisms and tools for technology co-operation in specific sectors;' ...

Though both the development and the transfer of technology are important aspects of the Bali Action Plan, transfer of technology is proving a particularly significant and complex issue in ongoing negotiations. Indeed, transfer of technology is taking an unprecedented place in the debate, given remaining technological disparities at the international level and the consideration of mitigation commitments for developing countries in a post-2012 climate regime. Significant divergences remain, moreover, as to the principal obstacles to the transfer of climate-related technologies, as well as to measures needed to effectively address these obstacles. Increased research and analysis will be fundamental in overcoming such differences, and in developing effective technology-related international co-operative action on climate change.

In any case, there will likely be no single answer to promoting the transfer of climate-related technologies. The UNFCCC has recognised that least developed countries, countries in the African region and small island developing states have 'specific needs and special situations' in regards to technology transfer[4]. Many of the tools found useful in other contexts, therefore, may be inadequate and even counterproductive for these countries, which have lower levels of development and other distinctive circumstances. For example, market-based technology transfer mechanisms such as foreign direct investment or joint ventures are ineffective in meeting the needs and demands of least developed countries[5].

The role of intellectual property (IP) in promoting the transfer of technology, though generally controversial, is particularly complex in relation to least developed countries. Studies show, for instance, that the basic conditions for IP to operate as an incentive for the transfer of technology do not exist in countries at the initial stages of technological development[6]. Moreover, unless they are selectively adapted to address the particular needs and conditions of least developed countries, IP rules may even hinder the technological development that is necessary for these countries to grow and reduce poverty, as well as address climate change mitigation and adaptation. As UNFCCC discussions address whether IP-related measures or other incentive mechanisms are necessary to ensure the transfer of the climate-related technologies, therefore, the singular situation of least developed countries should be taken into account.

Indeed, there is increasing realisation that – both within and beyond the intellectual property system – existing innovation structures and activities can and should be enhanced, and more efforts are needed to get smaller developing countries on board.

Given the complexity of the topic, the objective of this section is not to comprehensively address the topic of climate change, technology transfer and intellectual property. In the context of ongoing work on trade and climate change, however, it aims to provide an initial review of the links between these topics from the perspective of least developed countries, as well as to outline some of the relevant measures that could be developed in support of a post-Kyoto climate regime.

Technology transfer: role and impact of intellectual property rights

There is no single definition for 'transfer of technology'. On the contrary, both the term 'transfer of technology' and other related notions are understood in various ways depending on perspectives on technology and technology-related processes. In general, however, 'transfer of technology' can be defined as the transfer of systematic knowledge for the manufacture of a product, for the application of a process or for the rendering of a service[7]. The transfer of a technology is thus not exhausted in the transmission of the hardware, but also requires facilitating access to related technical and commercial information and the human skills needed to properly understand it and effectively use it[8]. These aspects are all recognised by the definition of 'transfer of technology' used in the UNFCCC context (see box 4.2).

> **Box 4.2** Intergovernmental Panel on Climate Change (IPCC) definition of 'transfer of technology'
>
> In its Special Report prepared in response to a request by the Subsidiary Body for Scientific and Technological Advice (SBSTA) to the UNFCCC, the IPCC defines 'technology transfer' as:
>
> > 'The broad set of processes covering the flows of knowledge, experience and equipment amongst different stakeholders such as governments, private sector entities, financial institutions, NGOs and research/educational institutions.
> >
> > 'The broad and inclusive term "transfer" encompasses diffusion of technologies and technology co-operation across and within countries. It comprises the process of learning to understand, utilise and replicate the technology, including the capacity to choose it and adapt it to local conditions'[9].

In the complex process of transfer of technology, the role of IP protection – despite being only one of many influential factors – has proven particularly contentious. IP rights were conceived as private rights to reward innovation and promote the dissemination of knowledge in the context of broader societal goals. By offering protection against a loss of control of information in technology-related transactions, IP is thus – in part – an instrument aimed at facilitating the transfer of technology[10]. On other hand, the existence of IP protection does not guarantee or suffice for effective transfer of technology. IP rights need to be buttressed by appropriate infrastructures, governance, and competition systems in order to be effective[11]. Moreover, there are circumstances in which IP rights may provide no incentives for the transfer of technology.

For the majority of least developed countries, IP protection may in fact hinder or prevent transfer of technology[12]. Research shows that, in the case of least developed countries, the absorption of existing technologies and their adaptation to local conditions primarily occurs through informal mechanisms, including imitation. In most cases, these informal mechanisms for the transfer of technology require some degree of 'reserve engineering', which can be curtailed by strong IP protection. In addition, the formal mechanisms for transfer of technology, including trade in goods, foreign direct investment, and licensing, which would be supported by stronger IP rights in other contexts, do not seem to have significant effect in least developed countries. Circumstances such as limited market size, weak regulatory mechanisms and minimal technological capability of local firms, for example, all limit the potential benefits from formal mechanisms for technology transfer in small and vulnerable economies[13].

The exact role of IP in the transfer of climate-related technologies remains unclear. In a 2006 UNFCCC report, IP-related issues did not feature prominently within a range of economic and market barriers to the transfer of technology in developing countries[14]. No comprehensive study has been conducted on the potential impact of IP rights in the different categories of climate-related technologies, but initial research found that the impact of patents on access to solar, wind and biofuel technologies in developing countries might not be significant[15]. On the other hand, studies by the European Patent Office (EPO) have noted the increasing number and scope of patent claims in wind energy and biofuels technologies – though the precise implications of these patent trends for the transfer of technology remain uncertain[16].

Nevertheless, the general dynamics of IP and transfer of technology are likely to remain valid in the climate change context. Determining a balance between the protection of IP rights and the promotion of public objectives, such as the transfer of technology, therefore, may be an important element of the post-Kyoto climate regime or the supportive international legal framework. Indeed, there are already significant calls to address the potential adverse effects of IP on the transfer of climate-related technology. On the eve of the Bali conference, for example, the European Parliament adopted a resolution, which stated that an ambitious post-Kyoto agreement might require 'corresponding adjustments' to be made to other international agreements, including on IP[17]. In discussions on the Bali Action Plan, moreover, several developing countries have stated as their position that IP is one of various obstacles that must be addressed in a systemic and cross-cutting manner to promote the transfer of technology[18].

Issues for developing countries

As discussion on the Bali Action Plan continues, recognising the special circumstances of least developed countries and their need for specific approaches and incentives for technology transfer should be paramount. UNFCCC Parties should thus consider a number of approaches that may ensure that IP protection functions as an instrument for the transfer of technology and other sustainable development goals in least developed countries. These approaches, which are non-mutually exclusive, include focusing on the policy space still available in international IP rules; increasing existing policy space through modifications to these rules; and developing new – and not necessarily IP-related – incentive mechanisms to promote the transfer of technology. Approaches focusing on protecting and enhancing the policy space for transfer of technology within the international IP system should also be considered.

The World Trade Organization (WTO) Agreement on Trade-related Aspects of Intellectual Property Rights (the 'TRIPS Agreement') introduced IP rights into the international trading system and remains the most comprehensive international agreement on the topic. The TRIPS Agreement states that the objective of the protection and enforcement of IP should be to contribute 'to the promotion of technological innovation and to the transfer and dissemination of technology, to the mutual advantage of producers and users of technological knowledge and in a manner conducive to social and economic welfare...'[19]. It also recognises that measures 'may be needed to prevent the abuse of intellectual property rights by rights holders or the resort to practices which ... adversely affect the international transfer of technology'[20].

Despite such language, it is unclear whether the TRIPS Agreement does establish a framework conducive to the transfer of technology, particularly in relation to least developed country Members. The TRIPS Agreement recognises 'the special needs and requirements of least developed country Members, their economic, financial and administrative constraints, and their need for flexibility to create a viable technological base', providing them with a special implementation period[21]. During this transition period, which is currently set to expire on 1 July 2013, these countries have available a range of channels for transfer of technology, including, for example, imitation and reverse engineering[22]. The TRIPS Agreement also foresees particular incentives for technology transfer to least developed countries (see box 4.3). Nevertheless, there are concerns as to how a strong IP regime, as encouraged and eventually required by the TRIPS Agreement, will impact the transfer of technology and other public policy considerations in least developed countries.

Box 4.3 Article 66.2 of the TRIPS Agreement

Article 66.2 of the TRIPS Agreement mandates developed country Members to 'provide incentives to enterprises and institutions in their territories for the purpose of promoting and encouraging technology transfer to least developed country Members in order to enable them to create a sound and viable technological base'.

Because least developed countries considered little progress was being made under this obligation, the Doha Ministerial Declaration mandated the TRIPS Council to 'put in place a mechanism for ensuring the monitoring and full implementation of the obligations' of Article 66.2. A decision setting up this mechanism was adopted in February 2003. It requires developed country Members to present annual reports providing a range of information on the incentives regime put in place to fulfil Article 66.2 obligations.

There are growing concerns, however, that such a mechanism is inadequate to promote effective transfer of technology in least developed countries. An analysis of reports on the measures taken to date, for example, found several shortcomings, including in the types and areas of incentives chosen[23]. As a result, it is still unclear to which degree transfer of technology takes place under Article 66.2 of the TRIPS Agreement, and what specific measures might be taken to improve the implementation of the provision.

This is where the issue of TRIPS flexibilities comes to the forefront of international discussions, including on the transfer of climate-related technologies. The TRIPS Agreement not only establishes minimum standards of IP protection, but also incorporates certain policy space to allow countries to position IP rights in the context of their broader objectives and priorities. For example, the TRIPS Agreement allows for certain limitations and exceptions to the protection of IP rights and for national determination of the appropriate method of implementation. These provisions are known as 'TRIPS flexibilities' and have been found to provide critical policy space in areas ranging from biodiversity and agriculture to public health and education. Though parallels with other public policy areas must be taken forth with care, previous experiences - particularly regarding patents and public health[24] - have become a reference point, and some UNFCCC Parties are already calling to reaffirm or increase TRIPS flexibilities to ensure a rapid and widespread transfer of climate-related technologies.

The TRIPS Agreement includes several types of provisions that could be useful in relation to climate-related technology transfer. For example, various provisions on patents - the exclusive rights granted for an invention, which generally constitutes a product or process that offers a new technical solution to a problem - are deemed pertinent to enhancing the transfer of technology. These provisions include:

- **Exemptions to patentability**: 'Patentability' refers to the boundaries established in relation to what inventions may be patented. Prior to the TRIPS Agreement, countries could exclude inventions of certain types or in certain areas of technology - such as pharmaceutical products and agricultural methods - from patentability, based on their development priorities and strategies. Article 27.1 of the TRIPS Agreement requires WTO Members to grant patents to all types of inventions in all fields of technology, as long as these inventions meet certain basic criteria. Because the TRIPS Agreement does not define the patentability criteria (namely novelty, inventive step and industrial applicability), however, some critical policy space remains in relation to the scope of patentability in each country. The loose definition of these criteria has raised concerns given the resulting all-encompassing patents. For example, patent claims on synthetic biology products and processes among the most promising technologies for cellulosic biofuels are so broad that scientists worry it could bring the discipline to a standstill[25]. Defining the patentability criteria to adequately limit the scope of patents, on the other hand, would have a positive impact on further innovation by limiting the possibility of conflict with existing patents. In addition, in some contexts, it would also enhance the transfer of technology. For example, by limiting the scope of patentability as they implement the TRIPS Agreement, least developed countries could safeguard informal pathways to the transfer of climate-related technologies, such as reverse engineering.

- **Exceptions to patent rights**: The TRIPS Agreement recognises that the rights of a patent owner to prevent third parties from exploiting the patented product are not absolute. Indeed, Article 30 states that WTO Members may provide 'limited exceptions' to these rights. That is, countries may - under certain circumstances - automatically allow the use of the patented invention by a third party without consent of the patent holder. The TRIPS Agreement does not define these circumstances, which will be linked to national policies and objectives. For example, a common exception addresses experimental use, allowing the use of patented inventions for research or experimental purposes by Parties other than the patent owner. This type of exception will be relevant in the climate change context, where adaptation of the technology to local needs and environments will be particularly vital. It would also allow companies in developing countries to 'invent around' patent claims to gain access to environmentally sound technologies, which has proved important in the context of the implementation of other multilateral environmental agreements.

- **Compulsory licences**: There are also other cases in which the TRIPS Agreement allows the use of a patented product or process without authorisation of the rights holder. One of the most important – yet perhaps most controversial – is the granting of compulsory licences. These non-voluntary licences are granted by an administrative or judicial authority to a third party, allowing the exploitation of the patented invention without consent of the patent owner[26]. Developing country Members consider this possibility as essential to ensuring that they can implement the TRIPS Agreement in a way that responds to broader public policies.

Article 31 of the TRIPS Agreement, which deals with compulsory licences, does not define the grounds on which countries may allow non-voluntary licences, although it does require a number of conditions and procedural steps. Climate mitigation or adaptation could provide valid grounds for compulsory licensing, and could even be considered to be included in general references to 'public interest' in most patent laws. Some countries also foresee compulsory licences in cases in which the invention is not exploited in the country, or is insufficiently exploited. Such a measure could restrain some of the anti-competitive practices feared as potentially impeding the transfer of climate-related technologies to developing countries. It is interesting to note that the issuing of compulsory licences is less arduous in certain situations, including cases of national emergency, other circumstances of extreme urgency or public non-commercial use. These compulsory licences could thus prove an effective tool to ensure rapid access to critical climate-related technologies in developing countries.

These examples show the potential contribution of TRIPS flexibilities to climate-related technology transfer, including for least developed countries once they are required or choose to establish IP protection. Moreover, increasing public attention and concerns about the relationship between IP and the transfer of climate-related technologies have resulted in calls for modifications to the TRIPS Agreements to support the post-Kyoto climate regime[27]. Proponents consider that changes such as explicitly incorporating climate protection as grounds for compulsory licensing or limiting the patentability of climate-related inventions could establish and consolidate policy space, which is important for a successful technology component in a post-Kyoto climate change regime[28].

Nevertheless, it is important to keep in mind the difficulties and vast political cost of modifications to the TRIPS Agreement. In addition, given the ongoing promotion of an agenda of higher levels and enhanced enforcement of IP protection, the risk of 'opening' the TRIPS Agreement should not be taken lightly. These points are particularly valid for least developed countries, which are currently not obliged to implement TRIPS Agreement provisions. Finally, on an issue as complex as climate-related technologies, it is questionable whether effective solutions could be achieved in the Council for TRIPS, a forum with a specific and limited approach. As a result, it is important to define the role of the UNFCCC and the climate regime itself in addressing the relationship between IP and climate-related technologies, including by developing new – and not necessarily IP-related – incentive mechanisms to promote the transfer of technology.

Intellectual property and the transfer of technology in the post-2012 climate regime

Both the UNFCCC and the Kyoto Protocol contain specific commitments on technology transfer. Article 4.5 of the UNFCCC urges developed country Parties, for example, to take all practicable steps to promote, facilitate and finance the transfer of, or access to, environmentally sound technologies and know-how, particularly to developing countries. Article 10 of the Kyoto Protocol, among others, reaffirms these commitments. Under Article 4.3 of the UNFCCC, moreover, developed country Parties are required to provide the financial resources needed by the developing country Parties to meet the agreed full incremental costs of implementing their obligations, including for the related transfer of technology. Indeed, the effective implementation by developed country Parties of their commitments on transfer of technology is inherently linked to the extent to which developing country Parties are required to implement their own commitments. As in other areas, however, realising the goals and complying with the obligations of transfer of technology have not proved simple.

The role of IP in promoting the transfer of climate-related technologies has traditionally not been at the forefront of discussions at the UNFCCC. As this debate appears to gain momentum, however, it is worth noting that UNFCCC Parties may consider a number of measures to enhance technology transfer in the context of a post-2012 climate regime – measures that may prove more feasible and effective than those sought in the context of the TRIPS Agreement. This is particularly true from the perspective of least developed countries. Some measures are already being discussed in the context of ongoing UNFCCC negotiations, including financial mechanisms to address the link between IP and the transfer of technology and guidelines on IP protection for publicly funded technologies. Other emerging topics include institutional arrangements for open or collaborative innovation, prizes as incentives to climate-related innovation and public-private partnerships.

Financial mechanisms are considered an important approach to addressing the issue of IP and transfer of technology. A 'Multilateral Technology Acquisition Fund', for example, has been proposed as a way to fund the transfer of technologies to developing countries through, inter alia, the buying-out of IP rights[29]. Given the relative success of the Multilateral Fund for the Implementation of the Montreal Protocol, such a proposal is actively being considered in the negotiations. Nevertheless, it should also be noted that, under the Montreal Protocol, 'IP rights did not constitute as large a barrier to technology transfer as was feared'[30]. Moreover, it is unclear that the case-by-case approach used in ozone-related technologies would work in the climate change context, given the greater range of relevant technologies and potential patent challenges[31]. On the other hand, complementary activities taken forth under the Montreal Protocol on technical assistance and capacity building set significant examples, particularly given the importance of enhancing national capabilities to achieve effective technology transfer in least developed countries.

Implications of public financing for the IP rights available over climate-related technologies has also been raised in the UNFCCC context, albeit not in recent negotiations. Government financing of research and development (R&D) – significant in most environmentally sound technologies – particularly benefits climate-related technologies. Nevertheless, such financing currently has few implications for the mode of ownership, commercialisation or transfer of these technologies, which are usually protected by IP rights[32]. As a result, these technologies, though stemming from publicly funded R&D, are not necessarily publicly available. A series of guidelines might guide public entities to retain some influence on the use and commercialisation of publicly financed climate-related technologies, and could be considered in the post-Kyoto climate regime.

As mentioned, there are other emerging measures that – while not discussed in the framework of the Bali Action Plan to date – would merit consideration. One is the use of open-source mechanisms to incentivise innovation and promote the transfer of climate-related technologies, as has been done in a variety of other fields, including software and biomedicine. Open source – which allows a broad access to the technologies developed beyond their innovators – has been called 'perhaps the most promising model' for technological development in least developed countries, given its costs advantages and the greater possibilities for learning due to the sharing of knowledge and ideas[33]. Its potential for the development and transfer of climate-related technologies remains largely unexplored.

Another possibility is the use of prizes to promote innovation while establishing a series of conditions that would allow the rapid and widespread transfer of the resulting technologies. In 2007, a number of prizes were offered by various countries and organisations to promote innovation in climate-friendly technologies, including the Earth Fund, launched in Bali to reward environmental innovation in areas such as second generation biofuels, water treatment or clean energies[34]. For these prizes to serve as an instrument of technology transfer, however, it is fundamental that related conditions are expressly considered and prioritised in their development.

Finally, another potential measure would be the introduction of public–private partnerships into the post-Kyoto climate regime. Such partnerships have been found to be particularly useful for transfer of technology to least developed countries, as they specifically aim to design and manage the various phases of the process successfully and might include specific arrangements on IP issues[35]. Moreover, in its Special Report on technology transfer, the IPCC noted that 'public–private partnerships are increasingly seen as an effective way in which the public sector can achieve public policy objectives by working with the private sector'[36].

Given remaining uncertainties, a unique conclusion on the relationship between IP and the transfer of climate-related technologies is far from feasible at the time of writing. Similarly, there is still little clarity as to the manner to best address this relationship in the various relevant international institutions and rules, particularly with relation to challenges posed by circumstances in least developed countries. Nevertheless, an overview of the potential opportunities and challenges presented by international IP rules to technology transfer under the post-2012 climate regime does present important lessons for possible next steps both in the UNFCCC and in the WTO.

First, it is clear that further research and analysis will be critical to achieve any effective solutions. An in-depth study of the various aspects of the interaction between IP and the transfer of climate-related technologies could provide the basis for more efficient and evidence-based discussions. Specific information on the climate-related technologies most strategic for least developed countries and the manner in which IP might impact the transfer of technology in practice could assist in moving negotiations towards more concrete problems and potential solutions.

Second, the existing transition period for the implementation of the TRIPS Agreement that is available to least developed countries provides a flexibility that should be noted. In addition, even when required to implement IP protection, least developed countries, as well as other developing countries, will still be able to resort to a number of TRIPS flexibilities to promote the transfer of climate-related technologies. The use of these flexibilities has not proved easy in other areas, but there is no evidence of such obstacles in the climate change context. Existing possibilities, therefore, should be explored in full. Third, it is important to note the need for negotiating expertise in the area of technology and IP rights – an expertise that is not shared by many environmental negotiators.

Finally, the consideration of measures related to IP and other incentive schemes should not be limited to the discussions on the TRIPS Agreement, but should also consider opportunities within climate negotiations. Considering IP-related issues in the development of measurable, verifiable indicators of technology transfer, for example, could help in ensuring adequate consideration of any positive and negative impact of IP on the implementation of the relevant UNFCCC and Kyoto Protocol provisions. In addition, a number of mechanisms increasingly explored and used in other public policy areas – including health, education and R&D – provide interesting examples to explore in ongoing UNFCCC negotiations as ways to enhance the technology transfer component of the post-Kyoto climate regime.

Moving beyond what has often been a general and political discussion, and – above all – moving towards an effective solution to the greatest development challenge of our time, requires such informed, comprehensive and coherent debate on intellectual property and climate change. It also requires the realisation that – in the context of least developed countries – technology transfer involves specific challenges and demands distinctive solutions.

Notes

1. IEA (2007a).
2. Steenblik, Vaughan and Waide (2006).
3. See, e.g. Article 4.1 (c) of the UNFCCC and Article 10 of the Kyoto Protocol.
4. The UNFCCC has recognised the particular needs and circumstances of these groups of countries in Article 4.9 and other subsequent decisions, including Decision 3/CP.13 on 'Development and transfer of technologies under the Subsidiary Body for Scientific and Technological Advice'.
5. Foray, D (2008).
6. United Nations Conference on Trade and Development (UNCTAD) (2007b).
7. Draft International Code of Conduct on the Transfer of Technology, 1985 version.
8. In this regard, a critical aspect of the technology transfer process is the development of the domestic capacities to absorb and master the received knowledge, innovate on that knowledge and commercialise the results.
9. Intergovernmental Panel on Climate Change (2000).
10. Studies have shown that such a positive impact does exist, including by establishing a link between stronger patent rights and productivity, trade flows, foreign direct investment and the sophistication of the technologies transferred. See, e.g. Maskus, K (2003).
11. Maskus, K. et al (2004).
12. United Nations Conference on Trade and Development (UNCTAD) (2007b).
13. Foray, D (2008).
14. UNFCCC Secretariat (2006b).
15. Barton, J (2007).
16. European Patent Office (EPO) (2007).
17. European Parliament resolution of 29 November 2007 on trade and climate change (2007/2003(INI)).
18. In the first meetings of the Ad Hoc Working Group on Further Commitments for Annex I Parties under the Kyoto Protocol (AWG) and Ad Hoc Working Group on Long-term Cooperative Action under the Convention (AWGLCA), Cuba, India, Tanzania, Indonesia, China and others stressed the need to address IP within technology discussions. Some developed countries, however, including Australia and the US, affirmed their belief that IP was not a barrier, but a catalyst for technology transfer.
19. TRIPS Agreement, Article 7.
20. TRIPS Agreement, Article 8.
21. TRIPS Agreement, Article 66.1.
22. These channels allow immediate and free access to some knowledge and facilitate the building of productive capacities in least developed countries. In addition, in 2002, WTO members

approved a decision extending until 2016 the transition period during which least developed countries do not have to provide IP protection for pharmaceuticals products.

23. Foray, D (2008).
24. Discussions on patents and public health at the WTO led to the adoption of the Doha Declaration on the TRIPS Agreement and Public Health, as well as an amendment to the Agreement to address the difficulties that WTO members with insufficient or no manufacturing capacities in the pharmaceutical sector could face in making effective use of some of the TRIPS flexibilities.
25. Suppan, S (2008).
26. The process does have a number of safeguards under Article 31 of the TRIPS Agreement, of course, including the requirements that the proposed user should have made good faith efforts to obtain authorisation from the patent holder, the use will be for domestic supply only, the patent holder shall be granted 'adequate' remuneration, that there be an established review process and so on.
27. In his speech to the UNFCCC Conference of the Parties in Bali, for example, the Brazilian foreign minister proposed that a statement similar to the Doha Declaration on the TRIPS Agreement and Public Health should be considered in the climate change context. The European Parliament, for its part, has recommended launching a study on amendments to the TRIPS Agreement required to allow for the compulsory licensing of environmentally necessary technologies.
28. Third World Network (2008).
29. See, for example, the statement of the African Group in COP-12 of the UNFCCC.
30. Andersen, S et al (2007).
31. In addition, it should be noted that in the case of ozone-depleting substances, alternative technologies to specifically and effectively address the problem had been identified and were available, which is not the case in the climate change context.
32. United Nations Conference on Trade and Development (UNCTAD) (1998).
33. United Nations Conference on Trade and Development (UNCTAD) (2007b).
34. Knowledge Ecology International (2008).
35. Foray, D (2008).
36. Intergovernmental Panel on Climate Change (2000).

5

Mitigation and Adaptation in Agriculture, Fisheries and Forestry

5.1 Climate Impact, Adaptation and Mitigation in the Agriculture Sector: Trade Linkages

Anthony Okon Nyong, International Development Research Centre
Nairobi, Kenya

Agriculture currently contributes significantly to the economy of many LDCs, SVEs and SIDS. On average, agriculture contributed about 2 per cent of GDP in developed countries in 2004, 11 per cent in developing countries and an average of 40 per cent in Africa (World Bank, 2007b). Globally, about 85 per cent of rural people derive their livelihoods from agriculture. In Africa, where more than 80 per cent of the population is rural, subsistence agriculture accounts for the livelihoods of about 90 per cent of this population, most of which live below official poverty lines.

The world is facing ever more severe environmental threats. These include erosion, land degradation, deforestation and climate variability and change, which could cause irreversible damage to terrestrial and marine ecosystems with consequent loss of production potential. About 40 per cent of the world's arable land is degraded, with most of the degraded soils found in the poorest countries. Here deforestation, over-exploitation and inappropriate land-use practices compound the problem. About 70 per cent of the cropland and 30 per cent of the pastureland in Africa are affected by degradation, and it is estimated that Africa accounts for about 30 per cent of the world's land degradation (UNEP, 2002).

Extreme climate events are a constant threat to agriculture and livelihoods in many developing countries. Since the 1970s, more frequent and intense droughts have been observed in the tropics and subtropics, where most of the world's poor reside. Increased drying linked with higher temperatures and decreased precipitation has exacerbated the drying up of already arid zones in Africa, in particular. Flows in many African rivers have declined by about 40 per cent since 1970, with grave implications for irrigation and energy generation (Niasse, 2005). As climatic patterns change, so also does the spatial distribution of agro-ecological zones. For instance, annual rainfall levels have been decreasing in the West African Sahel over the course of this century, with an increase in inter-annual and spatial variability. This has resulted in about a

200km southward shift in isohyets (Lebel et al., 1997; L'Hôte et al., 2002). Changing climate patterns have also resulted in changes in habitats, distribution patterns of plant diseases and pests, fish populations and ocean circulation patterns, all of which can have significant impacts on agriculture and food production.

In the low-latitude regions, where most developing countries are found, even moderate temperature increases are likely to result in declining yields for the major cereals. This could increase the risk of hunger in many parts of the world. More frequent extreme events like floods and droughts may lower long-term yields by directly damaging crops at specific developmental stages. Heavy rainfall could precipitate soil erosion, resulting in substantial agricultural loss. Several studies in Africa have established a positive relationship between drought and animal death.

Food security is a function of food availability, food access and food utilisation. Climate change could directly or indirectly impact on these three pillars of food security. It is estimated that in 2080 about 768 million people will be undernourished. Most of the undernourished will be in the developing countries, particularly in sub-Saharan Africa and South Asia, where crop production is projected to decline considerably.

At the same time, although globalisation offers opportunities for growth and development in all parts of the world, structural challenges faced by many developing countries have dimmed the hopes and promises that accompany the liberalisation of trade. Some of these challenges include low agricultural productivity, a heavy disease burden, particularly the HIV/AIDS pandemic with its debilitating effects on labour productivity, inflexible production and trade structures, low educational and skill capacity, poor infrastructure and deficient institutional and policy frameworks (FAO, 2007). These countries export a limited range of primary commodities, which are highly vulnerable to instability in supply, demand and a decline in terms of trade. In addition, the growing integration of markets has subjected their economies to a more fiercely competitive external trading environment. This is a level of competition they are neither prepared for nor equipped to deal with.

Globalisation has increasingly contributed to the marginalisation of the agricultural sector in many developing countries. For instance, the structural adjustment programmes of the 1980s, accompanied by complex market reforms, aggravated the vulnerability of many farmers in Africa. Fertiliser prices have risen in response to subsidy removal, resulting in largely negative responses to agricultural reforms. These have hindered the ability of many developing countries to improve their agriculture and to achieve the Millennium Development Goals (MDGs) through improving food security and increasing export earnings, benefits that were supposed to accrue from globalisation.

Global agricultural trade, and rules governing this trade, affect carbon management globally, as changes in land-use patterns have major impacts on the carbon balance. Agriculture is also the most vulnerable sector to climate change. This vulnerability is not only caused by the vagaries of climate, but also by its interaction with other developmental challenges. The impacts of climate change on trade will involve changes in comparative advantage based on environmental factors.

Both mitigation and adaptation measures can and should be taken within the agriculture sector. On the mitigation side, these include: cropland management, grazing land management and pasture improvement, restoration of degraded lands, livestock management, manure management and bioenergy. Many of these approaches provide 'win-win' outcomes or co-benefits in terms of higher productivity, better management of natural resources and habitats, or the production of valuable by-products. Others require substantial investment at the global level, such as the development of low-emission rice varieties and livestock breeds. The options are not universally applicable and smaller countries should evaluate and apply those best suited. In addition, possible negative effects or trade-offs of such measures should be considered.

Adaptation strategies include farm-level practices such as planting different crop varieties, changing planting dates, crop and livestock diversification, adapting to a shorter growing season, rotating or shifting production between crops and livestock, and shifting production away from marginal lands. Macro-level adaptation strategies include the introduction of insurance and other financial mechanisms to mitigate risks, as well as temporary migration where migrant farmers relocate from drought-affected areas to more favourable regions to farm, and subsequently return to their villages when conditions improve.

Adaptation to longer-term climate change is still in its infancy, despite 'showcase' examples of strategies. These strategies are often implemented as a planned adaptation and include the development and adoption of new technologies and strategies to reduce greenhouse gas emissions, as well as build the resilience of the agricultural sector. It is noteworthy that poorly designed and implemented adaptation strategies can result in maladaptation. This is where adaptive responses result in unintended, adverse, secondary consequences that outweigh the benefits of undertaking the strategy. There are also limits to adaptation, the most visible limit being finance. Some ecosystem species, human systems and cultures might be lost in the process of adapting.

The global community must support adaptation in the agriculture sector in developing countries. This also needs to be reflected in global trade negotiations. Trade justice should be promoted by ensuring that governments, particularly in poor countries, can choose the best solutions to end poverty and protect the environment. These may not always be free trade policies.

Climate change and trade policies will also impact significantly on other natural resources sectors of importance to LDCs, SVEs and SIDS, such as fisheries and forestry.

The climate problem

About one quarter of the CO_2 concentration in the atmosphere stems from agricultural sources (land-use change, deforestation and biomass burning); the remainder results from fossil fuel use. Emissions of methane amounted to 1,774 parts per billion (ppb) in 2005. While methane accounts for a smaller fraction of the warming effect than CO_2, most of it comes from agricultural sources such as domestic ruminants, forest fires, wetland rice cultivation and waste products. Nitrous oxide concentration was 319ppb in 2005. More than a third is primarily due to agriculture. Conventional tillage and conventional fertiliser use together account for 70 per cent of the N_2O. Altogether, agriculture is responsible for about 30 per cent of global warming. The largest share of the emissions of methane and nitrous oxide stem from developing countries, where agriculture plays a major role in livelihood strategies.

Agriculture is both a culprit and a victim of global warming. Efforts at achieving sustainable development should seek to reduce the contribution of agricultural emissions to climate change, as well as the vulnerabilities to and impacts of climate change on agriculture.

Impacts of climate change on agriculture

Agriculture is unarguably the most vulnerable sector to climate change. However, this vulnerability is not only caused by the vagaries of climate, but also by its interaction with other developmental challenges. This section draws substantially from the conclusions of the Fourth Assessment Report of the IPCC (2007b). The impacts of climate change on agriculture are discussed under biophysical impacts and socio-economic impacts, including food security.

Biophysical impacts

- *In mid- to high-latitude regions, moderate warming benefits cereal crop and pasture yields, but even slight warming decreases yields in seasonally dry and tropical regions*

  In mid- to high-latitude regions, temperature increases of between 1°C and 3°C across a range of CO_2 concentrations and rainfall regimes will likely have small but beneficial impacts on the main cereal crops – rice, wheat and maize. Further warming beyond this range will likely have increasing negative impacts. In the low-latitude regions, where most developing countries are found, even moderate temperature increases are likely to result in declining yields for the major cereals. This could increase the risk of hunger in many parts of the world. Simulations for sub-Saharan Africa estimate that countries such as Sudan, Nigeria, Somalia, Ethiopia, Zimbabwe and Chad, could lose cereal-production potential by 2080 across all emission scenarios (Fischer et al., 2005). However, global warming will also present opportunities for some countries to expand their agricultural potentials.

- *Increases in frequency of climate change extremes may lower agricultural productivity beyond the impacts of mean climate change*

  More frequent extreme events like floods and droughts may lower long-term yields by directly damaging crops at specific developmental stages. Heavy rainfall could precipitate soil erosion, resulting in substantial agricultural loss. Several studies in Africa have established a positive relationship between drought and animal death.

- *Impacts on weed and insect pests, diseases and animal health*

  Recent warming trends in Canada and the US have led to earlier spring activity of insects and the proliferation of some species, such as the mountain pine beetle. In Africa, rift valley fever epidemics, which always accompany El Nino events, could increase with a higher frequency of such events. This has a strong negative impact on human and livestock health on the continent.

- *Elevated levels of CO_2 and climate change will have varied impacts on livestock*

  Increased CO_2 and global warming will likely produce a dominance of unpalatable and invasive plant species, and could likely have detrimental effects on the nutritional value of extensive grasslands for grazing animals. In addition, increased temperature could lead to thermal stress, with a consequent reduction in animal productivity and conception rates. Increased climate variability will likely result in livestock loss. A temperature increase of up to 2°C could lead to increased net primary productivity in temperate regions, but in the semi-arid and Mediterranean regions, there will be no increase resulting in poor rangeland and fodder for livestock.

- *Changes in land availability for agriculture*

  Global warming will likely increase the area that is conducive to growth and production of agricultural crops, as well as extend the length of growing periods in some countries. For other countries, including those in Africa, a significant decrease in land suitable for rainfed agriculture is projected by 2080. In Africa, it is projected that the area of arid and semi-arid land could increase by 5–8 per cent. In southern Africa, climate change could expand the area that is likely to be unsuitable for agriculture by 11 per cent by 2080 (Fisher et al., 2005).

Socio-economic Impacts

- *Global cereal production and agricultural GDP*

  Most models generally agree that global cereal production would increase by as much as 200 per cent by 2080 with global warming because of CO_2 fertilisation (Fischer et al., 2005). More disaggregated regional models, however, have shown the disparity in cereal production at more localised levels. These detailed studies show an increasing gap in cereal production between developed and developing regions, especially after 2020. Whereas semi-arid developing countries, notably in sub-Saharan Africa and in some areas in South Asia where suitable arable land resources are limited, will see reductions in production in the range of 5–10 per cent, increases are projected for North America, Europe and the Russian Federation and parts of East Asia.

  The net loss in yield will exceed the net gains, and this is expected to cause a modest increase of between 2 and 20 per cent in the price of agricultural products in the short to medium term at the global level. Climate change could lead to an increase of about 2.6 per cent in agricultural GDP in some areas, particularly in high-latitude developed countries, and a reduction of about 1.5 per cent in others, particularly around the tropics, by 2080 (Fischer et al., 2005). Some parts of sub-Saharan Africa will experience a reduction of a magnitude greater than that predicted by global models. Most of Africa's agriculture is practiced on drylands and the area of semi-arid and arid lands in Africa could increase by about 60–90 million hectares (Boko et al., 2007). Model results show that dryland crop revenue falls an average of US$27 per hectare per 1°C increase in temperature (Kurukulasuriya et al., 2006). In South Africa, it is estimated that, without effective adaptation, crop net revenues will likely fall by 90 per cent by 2100 (Benhin, 2006). Livestock net revenue in the continent is estimated to fall by an average of US$379 per farm per 1°C. Considering the importance of agriculture to the economy of most African countries, climate change poses a real threat to development in the continent.

- *Food security*

  Food security is a function of food availability, food access and food utilisation. Climate change could directly or indirectly impact on these three pillars of food security. While the impacts of climate change on food availability and access are well known and documented, fewer studies have examined its impacts on food utilisation. It is estimated that by 2080 about 768 million people will be undernourished (Fischer et al., 2005). Most of the undernourished will be in the developing countries, particularly in sub-Saharan Africa and South Asia, where crop production is projected to decline considerably. A projected 2–3 per cent reduction in African cereal production by 2030 is enough to put 10 million people at risk of hunger. By 2080, the total population of today's more than 80 food-insecure countries is projected to increase from about 4.2 billion to about 6.8 billion. Some models project that 20–40 poor and food-insecure countries may loose on average 10–20 per cent of their cereal-production potential under climate change (Parry et al., 2004).

Climate change mitigation and adaptation

The United Nations Framework Convention on Climate Change (UNFCCC) has identified two human responses to global climate change and climate variability: mitigation and adaptation. Mitigation involves attempts to slow the process of global climate change by lowering the level of greenhouse gases in the atmosphere. Adaptation involves developing ways to protect people and places by reducing their vulnerability to climate impacts. Both actions are required to tackle climate change in the agricultural sector.

Mitigation

Agriculture contributes about half of the global emissions of two of the most potent non-carbon dioxide greenhouse gases – nitrous oxide and methane – and about 30 per cent of total greenhouse gas (GHG) emissions. Nitrous oxide emissions from soils (from fertiliser application and manures) and methane from livestock production are projected to rise with climate change. The remainder of non-carbon dioxide emissions are from biomass burning, rice production and manure management. Agriculture also contributes to reduce carbon sequestration through deforestation and land-use change. Agriculture therefore offers excellent opportunities for reducing GHGs. Given that most of the agricultural contribution to GHGs is from the developing countries, about 70 per cent of the economic potential for mitigation is also in those countries.

Mitigation options

Sustainably mitigating climate change requires identifying effective ways to reduce greenhouse gases produced and released to the atmosphere. Various mitigation options have been considered in the agricultural sector. These include: cropland management, grazing land management and pasture improvement, management of organic/peaty soils, restoration of degraded lands, livestock management, manure management and bioenergy. Many of these approaches have win-win outcomes in terms of higher productivity, better management of natural resources or the production of valuable by-products, such as bioenergy. Others require substantial investment at the global level, such as the development of low-emission rice varieties and livestock breeds.

Considering that about 90 per cent of mitigation potential in agriculture is through better cropland management, many existing practices address this strategy. Some of these practices include:

- Agronomy: Implementation of agronomic practices that give higher yields and residues, which can increase soil carbon. It also involves the development of better crop varieties, and use of cropping rotations and cover crops.
- Nutrient management: This involves improving the efficiency of nitrogen use. Since fertiliser is responsible for large amounts of nitrous oxide emissions in the agricultural sector, farmers can be encouraged to choose management practices that lead to appropriate fertiliser application rates.
- Tillage/residue management: Practices include low- to zero-tillage crop management practices, and the retention of residues on farms.
- Water management: The improvement of yields through good and efficient irrigation practices, and better water management.
- Rice management: Rice production is responsible for a large proportion of methane emissions from agriculture. These emissions are generated through the cultivation of paddy rice, which promotes the anaerobic decomposition of plant wastes that remain after the harvest. Reductions in methane emissions can be achieved by cultivating low CH_4 emitting varieties, improving water management practices and using inorganic fertilisers.
- Agro-forestry: Establishment of shelter belts and riparian buffer strips with woody species.

- Land cover (use) change: Converting cropland to grassland; reversion of cropland to natural vegetation.

However, it must be noted that these options are not universally applicable – individual countries should evaluate their circumstances and their applicability based on their land use and management, and climate and social settings. They must also consider uncertainties that may be associated with these options, such as the level of adoption and potential barriers to such adoption, the technical effectiveness of each of the options and the persistence of mitigation.

Mitigation trade-offs

It is obvious that some mitigation practices that benefit agriculture can have adverse impacts in other sectors. For instance, draining the land under rice cultivation can adversely affect wildlife habitat, reduce wetland acreage, as well as reduce wildlife benefits. There are also mitigation practices that have co-benefits. For example, reducing nitrogen fertiliser use can improve water quality and reduce nitrous emissions. In addition, providing riparian buffers can enhance wildlife habitat, improve water quality and increase carbon storage. It is therefore important to consider trade-offs in implementing any mitigation strategy.

Adaptation

Adaptation is the adjustment in natural or human systems in response to actual or expected climatic stimuli or their effects, which moderates harm or exploits beneficial opportunities. Adaptation to climate change should address both short-term climate variability and long-term climate change. While the impacts of climate variability are essentially local in scale, climate change will affect long-term trends in agricultural development. The following paragraphs review various adaptation strategies in the agricultural sector that address both the short- and long-term impacts.

Adaptation to short-term climate variability

Adaptation to climate variability in the agricultural sector is not a new phenomenon. Several strategies are being implemented. The first involves management practices at the farm level. Of these, the ones that address crop vulnerability include planting different crop varieties, changing planting dates, crop and livestock diversification, changing land-use practices, adapting practices to a shorter growing season, rotating or shifting production between crops and livestock and shifting production away from marginal lands. Short-term adaptation strategies specific to livestock management include shifts in biological diversity, species composition and/or distribution; changes in grazing management (timing, duration and location) and varying supplemental feeding; changing the location of watering points; altering the breeding management programme; changes in rangeland management practices; and modifying operation production strategies, as well as changing market strategies (Kurukulasuriya and Rosenthal, 2003).

A second strategy involves reliance on insurance and other financial mechanisms. Agricultural insurance is widely recognised as a veritable strategy for spreading risks, including climate-induced risks. It can enhance financial resilience to external shocks and provide a unique opportunity to spread and transfer risk. Insurance, if properly implemented, may provide incentives for risk reduction and prevention, while engaging the private sector in climate change response action.

However, the insurance market is currently limited in developing countries, in part due to limited economic assets, limited private sector interest in insurance and the weak financial institutions in these countries. In addition, few of these countries have any detailed quantification or economic quantifications of their climate risks, and as such cannot pursue the option of insurance as they do not know what the premium would be. Consequently, there is a greater dependence on informal risk coping strategies, such as risk pooling and micro-credit facilities.

Given that so few developing countries have detailed quantifications of the economic valuations of their risks to climate change, one of the benefits of promoting insurance-related actions is that it may help advance efforts on quantifying risks and potential losses due to climate change (UNFCCC, 2007a). Minimising risk can then result in a reduction of the rates for insurance, which thereby become more affordable.

A third short-term adaptation strategy is temporary migration, where migrant farmers relocate from drought-affected areas to more favourable regions to farm and subsequently return to their villages when conditions improve. Sometimes families agree to send family members to work in other regions or countries, and these individuals in turn support the family through remittances. These remittances have often contributed significantly to providing the needed credit to buy farm inputs and implements to improve agricultural productivity.

Adaptation to long-term climate change

Adaptation to longer-term climate change is still in its infancy, despite 'showcase' examples of strategies. These strategies are often implemented as a planned adaptation and include the development and adoption of new technologies and strategies to reduce greenhouse gas emissions, as well as to build the resilience of the agricultural sector. Such technologies include soft technologies such as information systems, management practices, development of new crop cultivars etc., and also hard technologies such as development of equipment for irrigation, conservation tillage and integrated drainage systems. Adaptation strategies are often site-specific and their implementation should reflect numerous decision rules. For instance, such rules should include considerations of the extent of belief that the climate is actually changing, awareness of the type and form of change, knowledge of technology, not only today but also in years to come, and assumptions about what governmental policies will be in various regions and over time.

Yet mitigation is still the best adaptation strategy, and some developed countries have recognised this and have set aggressive targets to reduce greenhouse gas emissions.

Maladaptation and limits to adaptation

Poorly designed and implemented adaptation strategies can result in maladaptation. This is where adaptive responses result in unintended, adverse, secondary consequences that outweigh the benefits of undertaking the strategy. An adaptive response that is made without consideration for interdependent systems may, inadvertently, increase risks to other systems that are sensitive to climate change. However, even when a comprehensive approach is taken to the development of strategies for adapting to climate-induced effects, one must account for potential non-climate related side effects of the adaptive strategies to avoid maladaptation. The social acceptability of a particular adaptive response may depend upon who in society will benefit from the adaptation policy and who will lose (Smith et al., 2007). There are also numerous examples where specific adaptations impose adverse social externalities (not all adaptation is always good for everyone). For instance, the building of a dam for irrigation to improve agricultural yield could also create suitable habitats for disease vectors. There are also instances where adaptations to moderate risks actually enhance vulnerability to larger events.

At the same time, there are limits to adaptation, the most visible of which is finance. Current estimates of costs are tentative and depend on the climate change scenario and how ambitious the adaptation regimes are expected to be. However, some studies have estimated the total cost of adaptation in agriculture to be about US$50–86 billion each year (Raworth, 2007; UNDP, 2007).

Besides finance, other limits are perhaps even more critical. Even with all the money in the world, it is still not possible to guarantee complete adaptation. There are systemic limits to adaptation where some ecosystem species, human systems and cultures might be lost in the process of adapting. The higher the rate and/or magnitude of climate change, the more the systems that will be 'adapted out of existence'.

There are also informational limits. The climate variables and spatial scales with the *most relevance* for adaptation decisions in the agricultural sector are generally the ones projected with the *least certainty* (e.g. precipitation, extreme weather, local-scale events etc). Understanding the limits to adaptation will enable us think of more integrated strategies to reduce the impacts of climate change.

Virtual water and ghost acres

Global models are not unanimous in predicting how global warming will influence the magnitude and patterns of precipitation. However, it is generally agreed that many areas will experience a drying that will negatively affect the hydrological cycle. Many governments are becoming increasingly interested in where their limited water supplies are going and what they are being used for. This is with the view to better manage their water resources to adapt to climate change effectively. This raises the question of 'virtual water' and 'ghost acres'.

'Virtual water' is calculated in terms of water that is used to grow crops that are exported to or imported by other countries. So, water that is used to grow flowers in

Kenya, for example, is actually serving as a supplemental water supply for other countries that import the flowers; that is, as an opportunity cost for the exporting countries, as this water is not being used to generate energy or produce food in those exporting countries. It is estimated that annually, the UK 'imports' 189 million m^3 of African water through the import of green beans. This is enough to provide 10 million Kenyans with drinking water (McCulloch and Ota, 2002). Given that Kenya is categorised as a

Box 5.1 Biofuels

With rises in oil prices and with few alternative fuels for transport, coupled with a desire to reduce the dependence on fossil fuels with their large CO_2 emitting capabilities, several countries are actively supporting the production of liquid biofuels from agriculture. Biofuels offer a potential source of renewable energy and could lead to large new markets for agricultural producers. The increased production of liquid biofuels has led to a corresponding increase in the debate on its desirability, drawing attention to the potential positive and negative impacts of the product. While proponents of biofuels point to the potential for cleaner fuels, greater economic opportunities for farmers and rural communities and a renewable source of energy, opponents argue that biofuels risk damaging biodiversity resources, marginalising local communities and indigenous groups and creating more greenhouse gas emissions than they prevent.

This debate is complicated by the fact that numerous types of feedstocks can be used in the production of biofuels. Depending on the feedstock used, where and how it is grown and the manner in which it is processed, the greenhouse gas balance, energy yields and environmental impacts of biofuels may differ greatly (The Royal Society, 2007). In addition, depending on what assumptions and methods are used in estimating the impact of biofuels, the magnitude of the potential impacts can vary significantly. Further complicating the situation is the fact that biofuel technology and policies are evolving at a rapid pace. Given this complexity, it is not possible to generalise the specific impacts of biofuels as each fuel type and system of production has different potential impacts. However, because of the ethical issues of 'food versus fuel', it is important that developing countries contemplating embarking on massive biofuel production should do so cautiously after carrying out thorough cost-benefit and impact analyses of the entire production life cycle.

Several studies emphasise that the production of biofuel crops could have a negative impact on water resources, especially when traditional field crops are used (De Fraiture et al., 2007; Berndes, 2002). Certain crops such as oil palm, sugar cane and maize can have very high water requirements, which lower their water-use efficiency (Steenblik, 2007; Rajagopal and Zilberman, 2007). The production of biofuel feedstocks could also have a variety of positive and negative impacts on socio-economic conditions. As the majority of feedstocks used in the production of biofuels are agricultural, the market for biofuels and agricultural products are closely related (Doornbosch and Steenblik, 2007). The rising demand for agricultural biofuels is translating into higher market prices for some agricultural products. Biofuels, by using agricultural crops for feedstocks, reduce the amount of crops and land available for food production, resulting in higher commodity prices even into the next decade (UNCTAD, 2007c). The clearest example is maize, the price of which rose by over 60 per cent from 2005 to 2007, largely because of the US ethanol programme combined with reduced stocks in major exporting countries. The price of all crops used in the production of biofuels, with the exception of sugar, has increased dramatically over the same period of time.

water-stressed country, and this is forecast to worsen, the implications for an expansion of the green bean trade could raise fundamental issues when considering adaptation to climate change in the water sector. While there are no clear guides on how governments should deal with the issue of virtual water, they should evaluate the issue in terms of climate change and related food balance equations, and in terms of its water and food security situations.

'Ghost acres' refers to the amount of land that is used to cultivate food that is exported to support other countries. Like virtual water, ghost acres should also be taken into consideration in the food security equation and in the climate change debate. The debate on ghost acres is also gathering momentum, as developing countries are becoming increasingly concerned that climate change could reduce the overall amount available to produce food to feed their ever-growing populations.

Implications of climate change for international agricultural trade

Agricultural trade and climate change linkages

The links between climate change and international agricultural trade and markets are obvious. The supply of agricultural produces within a country is a function of the volumes produced domestically, the price of imports and the price of the exports used to generate foreign exchange. Climate change could affect all three variables. Previous sections have shown how climate change could affect agricultural production directly.

While many countries do strive to maintain food self-sufficiency, trade enables more people to have greater access to food in times of shortages. Countries that experience shortfalls in periods of insufficient rain can import food, while countries that are less affected can sell their produce in areas more severely affected by climate change.

Climate change can have a direct or indirect effect on the price of agricultural imports. Climate models have shown that increased production from the areas that will 'benefit' from climate change will be smaller than the decline in those that will 'lose' global supply. This will most likely lead to a rise in agricultural prices. This direct impact is additional to the other influences on global demand and supply (such as rising incomes, increased population and hence greater demand and an increased drive for production of biofuels).

Some climate change mitigation strategies will also have implications for agricultural prices. In an effort to reduce consumption of fossil fuels and hence cut greenhouse gas emissions, there is a significant global increase in the production of biofuels (see box 5.1). This has contributed to the current food crisis by raising agricultural prices as a result of diverting agricultural resources away from food production.

Many developing countries generate a large part of the foreign exchange they require to fund imports through the export of agricultural goods. In sub-Saharan Africa, this forms 50 per cent of the region's export earnings. The net impact of climate change will result from a combination of the impact on their production of exported crops and demand for these in the world. Policies to reduce food miles could raise the price of vegetables and flowers in import countries while causing a glut in export countries, thereby reducing the ability of many developing countries to earn foreign exchange to participate in international trade.

At the same time, climate change can directly affect trade-related infrastructure or trading routes. For instance, rising sea levels may endanger coastal infrastructure that supports trade, such as ports. In addition, extreme weather events can be expected to disrupt markets and infrastructure. One of the predicted effects – increased flooding – will affect infrastructure as well as transport routes, as was the case during the 2000/2001 flooding in Mozambique when roads and railway lines were washed away. This is against the backdrop that Africa's road infrastructure is worse than any other continent. As noted by Sachs et al. (2004), before high-intensity modern trade can get started, Africa needs an extensive road system both from the coast to the interior and within the interior, where the highest population concentrations are found. These roads, however, are very expensive to build and maintain. Destruction of existing roads as a consequence of climate change will severely affect trade in Africa.

Climate change and agricultural trade policies

Trade liberalisation: Trade liberalisation is expected to lead to improvements in access to international markets, which in turn can help a country diversify and reduce the risk of food shortages from climate change. This argument is based on the assumption that the trading system is a risk-spreading mechanism, working through the geographic relocation of world food supplies according to changing comparative advantage and spatial diversification of climatic risks.

However, it is important to note that trade policy of the WTO might serve to facilitate or hinder efforts to address climate change mitigation and adaptation. For instance, while trade liberalisation would encourage an increase in the scale of production to reduce transaction and production costs, the scale effect will, in and of itself, have negative climate change impacts; the more goods and services produced under present technologies, the more greenhouse gases (GHGs) emitted. Increased trade through liberalisation will obviously lead directly to more global GHG emissions from increased transport of goods. The GHG-intensity of transport varies enormously from marine transport to trucks to airfreight, but ultimately all modes of transport have some emissions. At the same time, trade liberalisation could have beneficial impacts on mitigation strategies where investment agreements may bring new techniques of production that are more energy efficient, and therefore emit fewer GHGs per unit of output. This may be due to foreign investors bringing new technologies, or domestic firms having to increase efficiencies in the face of foreign competition.

Tariffs: Another specific WTO trade policy that might affect efforts to combat climate change is the tariff structure. It is generally argued that high tariffs on imported goods and other trade-restrictive policies can reduce the effectiveness of trade, as well as impede the entry of efficient technologies into new markets. This has implications for technology transfer to combat climate change. This has led to calls for reductions in agricultural tariffs and subsidies, noting that the removal of distortions in global agricultural activities is likely to improve allocative efficiency in agriculture and improve aggregate welfare if it is accompanied by the removal of farm-support mechanisms. Others have argued that there should be selective tariff lowering, if at all, as lower tariffs can aggravate climate change impacts. For instance, lowering tariffs on paddy rice is likely to aggravate climate change through increased methane production.

Climate change, comparative advantage and competitiveness

Climate change will have significant impacts on agricultural trade flows. In general, the impacts of this type will involve changes in comparative advantage based on environmental factors. Kenya's role as a tea-exporting powerhouse, for example, may be under threat from climate change-related drought in the long term. The same applies to other products and countries. A country that has a comparative advantage today in the production of a particular agricultural commodity may lose that advantage in the future due to climate change threats.

Implications for sustainable development

For a long time, climate change has been viewed largely as an environmental issue of little relevance to development. For the same reason, development approaches have not been given the necessary attention by the climate change community, who instead focused on reducing greenhouse gas emissions. This state of affairs is partly to blame for the heavy price countries continue to pay whenever climate disasters strike and resources are diverted to attend to such events, instead of having long-term strategies that may not only minimise the impacts but also ensure resources meant for development are not diverted to deal with such emergencies.

Climate change can potentially undermine whatever modest gains have been achieved towards meeting the Millennium Development Goals (MDGs). It is therefore important that climate change be mainstreamed into developmental policies and plans. While climate change will likely affect development at various levels, the development approach chosen will also influence future emission of greenhouse gases as well as the adaptive capacity of individuals, communities and countries. According to Huq et al. (2006), unsustainable development is the underlying cause of climate change and the development path taken determines the degree to which society is vulnerable to climate change.

Poverty is a strong impediment to achieving sustainable development and agriculture holds the key to reducing poverty in many developing countries. The large share of agriculture in GDP in low-income countries suggests that strong growth in agriculture is necessary for overall economic growth. Indeed, agriculture has accounted for about one-third of growth in sub-Saharan Africa over the past 15 years. As GDP per capita rises, agriculture's share in GDP declines, and so does its contribution to growth.

Unfortunately, agriculture appears to be the most vulnerable sector to the adverse impacts of climate change. To achieve sustainable development, efforts need to be stepped up to mitigate and adapt; both strategies are very important and should be pursued concurrently. Developed countries need to actively pursue strategies to reduce greenhouse gas emissions. Even if greenhouse gases were stabilised in the atmosphere today, global warming would continue for a long time.

Climate change requires a global framework for international co-operation, a vital part of which is adaptation action. Actions to enable adaptation to climate change also pose opportunities to promote sustainable development. Yet developing countries require resources in order to promote these actions. A successful framework must directly involve assistance for adaptation in developing countries, particularly small island developing states and least developed countries, given that they will disproportionately bear the brunt of climate change impacts. Combating climate change is vital to the pursuit of sustainable development; equally, the pursuit of sustainable development is integral to lasting climate change mitigation.

Need for global collaboration

There is a lot that the international community can and needs to do with respect to climate change. First, the UNFCCC tasks Annex I countries to support non-Annex I countries to reduce their vulnerability to climate change. This requirement goes beyond just providing finances as aid to the developing countries. Annex I countries have a moral responsibility to substantially reduce their greenhouse emissions to a level that will reduce the extra burden that climate change poses to poor developing countries. While many of these countries have signed the Kyoto protocol, they are yet to have clear-cut plans and strategies to meet the commitments of that protocol. Just as the world, concerned with the state of poverty in the developing countries, came up with the MDGs (measurable and achievable targets), the same needs to be done regarding green house gas emissions.

The international community should also support adaptation in developing countries. One reason for Africa's high vulnerability to climate change is its low adaptive capacity, suggesting that efforts should be made to improve this. Adaptation is already considered a vital part of any future climate change regime. Within the UNFCCC and the international community, deliberations are building to find an effective means to tackle climate change. Future decisions within the UNFCCC negotiating process must assist developing countries in a streamlined, innovative and transparent way, with transfer of knowledge, technology and financial resources to adapt at all levels and in all sectors. If there are delays in implementing adaptation in developing countries, including delays in financing adaptation projects, this will lead ultimately to increased costs. Delays in implementing adaptation will also lead to greater dangers to more people. For example, extreme events including droughts and floods could trigger large-scale population movements and large-scale conflict due to competition over scarcer resources such as water, food and energy.

The international community can further support the development and transfer of low-carbon intensive technologies to developing countries. Most developing countries are in a position to chart a path towards green development, without following the model of the present industrialised nations. However, this requires financial and technical assistance from developed countries.

The international community can further increase research funding for Africa. Despite several efforts, global research into areas critical to African development remains woefully under-funded. For example, the annual operating budget of US$400 million for the worldwide network of 16 tropical agricultural research centres known as the Consultative Group on International Agricultural Research (CGIAR) is minuscule compared with the research and development (R&D) budgets of the world's six largest agrobiotech companies, estimated at roughly US$3 billion a year (Sachs et al., 2004). The budget of the CGIAR system, as well as of national agricultural research centres, remains low despite considerable evidence of the high social rates of return from R&D in tropical food production.

There is the need to reform agricultural trade and increase developing countries' participation in international trade more generally. For instance, trade discrimination against Africa has persisted for decades, despite repeated promises to remove limits on the continent's access to US and European markets. Even new initiatives to expand access to the US market, such as the African Growth and Opportunity Act, remain heavily constrained by rules of origin and other limits to African exports. It is important to ensure that subsidised crops from the North and market protection do not undermine the efforts of farmers in Africa to sustain their livelihoods. Trade justice should be promoted by ensuring that governments, particularly in poor countries, can choose the best solutions to end poverty and protect the environment. These may not always be free trade policies. Export subsidies that damage the livelihoods of poor communities around the world may need to be abolished.

Agriculture holds the key to economic growth in the developing world, yet the sector will be severely impacted by climate change. There is the need to reduce greenhouse gas emissions as well as adapt to the adverse impacts of climate change. Given that agriculture is both a culprit and a victim of global warming, properly managing agricultural practices could contribute substantially to climate change mitigation. Meanwhile, efforts are increasing to get developing countries to adapt to the adverse impacts of climate change. There are, however, limits to adaptation. Nor should the potential for adaptation lead to complacency. Agricultural adaptation to climatic variation is not perfect, and changes in how farmers operate or in what they produce may cause significant disruption for people in rural regions. Indeed, some adaptive measures may have detrimental impacts of their own.

Many developing countries have a good opportunity to chart developmental pathways that are more sustainable. Developed countries can also change their developmental paths to such as will foster climate change mitigation. All countries should choose and implement mitigation and adaptation options that will bring about synergies and avoid conflicts with other dimensions of sustainable development. Implementation of these mitigation strategies will require resources to overcome multiple barriers, particularly for developing countries. It is therefore important to have a strong global collaboration to address climate change impacts in the developing countries. While the UNFCCC tasks developed countries to assist the developing countries to combat the adverse effects of climate change, developing countries can also play a major role by mainstreaming climate change into their normal developmental policies and plans.

While trade liberalisation holds a lot of promise for agricultural development, a completely liberalised market will not bring the same benefits to all countries. It may cause difficulties for some developing countries, in particular those that are dependent upon food imports or those losing preferential access to markets. Measures will be needed to help such countries adjust. Nevertheless, in the long term, liberalisation should result in a more favourable international structure of agricultural prices, which should benefit most poor countries.

The most pressing challenge is to address the environmental and economic challenges of the poorest nations to reap the benefits of agricultural production in the bid to achieve sustainable development and the MDGs. There is the need for trade and climate change policies to be streamlined so that developing countries that will be more adversely affected by climate change would not see their vulnerabilities exacerbated by poor and restrictive trade policies.

5.2 Climate Change and Fisheries: Policy, Trade and Sustainable Development Issues

Dr Moustapha Kamal Gueye, ICTSD

Fisheries provide more than 2.6 billion people with at least 20 per cent of their average per capita animal protein intake. In 2004, an estimated 41 million people worked as fishers and fish farmers, the great majority of these in developing countries. Thirty-eight per cent of total fisheries and aquaculture production was exported, amounting to an export value of US$71.5 billion. Net fishery exports by developing countries grew from US$4.6 billion in 1984 to US$16.0 billion in 1994 to US$20.4 billion in 2004. There has been a doubling of the share of fish trade in both total GDP and agricultural GDP over the past 25 years (FAO, 2006).

Despite its vital socio-economic importance, overexploitation and risk of stock depletion remains a pressing concern in the fisheries sector. Today, 25 per cent of marine fish stock are overexploited and depleted; more than 50 per cent are fully exploited. Most of the stocks of the top 10 marine species, which account in total for about 30 per cent of the world capture fisheries production in terms of quantity, are fully exploited or overexploited. This confirms earlier observations that the maximum wild capture fishery potential from the world's oceans has probably been reached, and reinforces calls for more cautious and effective fisheries management to rebuild depleted stocks and prevent the decline of those being exploited at or close to their maximum potential (FAO, 2006).

Climate impact on fisheries

The critical situation of many marine fisheries is likely to be compound by the effects of climate change. The Fourth Assessment Report of the Intergovernmental Panel on Climate Change (IPCC, 2007b) foresees impacts of climate variability and change on fisheries in coastal and estuarine waters, although non-climatic factors, such as overfishing and habitat loss and degradation, are mainly responsible for reducing fish stocks.

Several countries, including a large number of developing and least developed countries (LDCs) that depend on fisheries for food security, employment and government revenues, present a high degree of vulnerability to climate change. Such vulnerability has been measured as a factor of risk exposure, sensitivity and adaptive capacity (see table 5.1). The impacts of climate on the various fisheries resources are not yet fully understood. Such impacts are likely to be different depending on the region, the species and the state of the stocks – and can be either positive or negative. For example, some research indicates that future climate change impacts will be greater on coastal than on pelagic species, and for temperate endemics than for tropical species. It will

Table 5.1 Projected indices of vulnerability of fisheries to climate change

Rapid development, high emissions scenario (IPCC A1F1)		*Local development, lower emissions (IPCC B2)*	
Angola	81.97	Mauritania	83.10
Mauritania	81.18	Angola	82.15
Niger	79.24	Zimbabwe	79.32
Dem. Rep. of Congo	78.82	Niger	78.95
Mali	78.01	Dem. Rep. of Congo	76.03
Sierra Leone	77.09	Mali	75.92
Burkina Faso	76.01	Mozambique	75.13
Burundi	74.96	Russian Federation	74.33
Mozambique	74.86	Sierra Leone	73.61
Zimbabwe	74.55	Senegal	73.31
Senegal	73.70	Botswana	72.96
Guinea Bissau	72.97	Zambia	71.78
Côte d'Ivoire	71.18	Burundi	71.68
Sudan	70.68	Burkina Faso	71.57
Russian Federation	70.57	Peru	70.98

Source: Allison et al. (2005); Composite index of vulnerability = risk exposure + sensitivity + adaptive capacity[1].

also lead to migration of species in certain regions (Brander et al., 2003). With higher temperatures, increased invasion by non-native species is expected to occur, particularly on mid- and high-latitude islands. All these changes may alter current patterns of production and trade in fisheries, with potential gains in certain regions and loss in others.

Issues for developing countries

Climate change will increase uncertainties in the supply of fish from capture and culture. Such uncertainty will impose new challenges for risk assessment, making it critical for resources planning and management to take into account the greater possibility of unforeseen events, such as the increasing frequency of extreme weather events and 'surprises'. A recent meeting of experts on fisheries, climate change and food security has concluded that in the face of these impacts and the existing development and management constraints, the primary challenge for the sector will be to deliver food supply, strengthen economic output and maintain and enhance food security while ensuring ecosystem resilience[2]. The long-term productivity of stocks is related to the carrying capacity of their environment. These may alter with overfishing, as well as changes induced by climate change.

The future implications of climate change make it urgent to advance policy reform in capture fisheries, including addressing the problem of overcapacity and overfishing as part of mitigation and adaptation strategies. A number of mitigation and adaptation measures have been identified including:

- Estimating future production levels, taking into account environmental variability and the impact that this may have on the growth of a stock and on its reproductive and mortality rates,
- Developing tools for decision-making under management uncertainties and expand societal knowledge,
- Adjusting fleet and infrastructure capacity and flexibility, controlling fishing effort and enhancing management systems while creating alternative employment and livelihood opportunities,
- Expanding aquaculture along sustainable and equitable development paths with appropriate legal and regulatory frameworks,
- Enhancing emergency preparedness and response, and developing insurance and social safety schemes in the fisheries sectors,
- Applying international fishery agreements and conventions more vigorously, and strengthening these if necessary, to accommodate and support climate change-related activities,
- Integrating climate considerations into investment in the sector, especially in infrastructure, and

- Adopting measures to reduce the carbon footprint of the fisheries sector, and providing payment for environmental services, particularly those offering additional livelihood options to poorer communities.

Responding to climate change in the fisheries sector may have crucial implications for the ongoing effort at the World Trade Organization to discipline fisheries subsidies that lead to overcapacity and overfishing, taking into account the importance of this sector to development priorities, poverty reduction and livelihood and food security concerns in developing countries. As WTO negotiations enter a crucial stage, climate considerations have not surfaced in the talks. However, the Doha outcomes on fisheries may have important bearings on the sector, as curbing overfishing can lead to healthier marine fisheries stocks, thereby increasing resilience and capacity to adapt to the impact of climate change.

Box 5.2 The forestry sector and trade in smaller developing countries: the potential of Reduced Emissions from Deforestation and Forest Degradation (REDD)

Avoided deforestation and bio-sequestration (through afforestation or reforestation activities) are well-established ways to address climate change. Currently, deforestation continues apace with an annual global loss of 16 million hectares. Central among the economic drivers of deforestation include unsustainable demand from industrialised countries for timber products, the expansion of agricultural production and the failure of conventional economics to recognise the full economic value of the natural capital and ecosystem services provided by forests.

Maintaining forests as carbon sinks will make a significant contribution to stabilising atmospheric greenhouse gas concentrations. Forests play a particularly critical role in the mitigation of global climate change through carbon sequestration, the maintenance of hydrological cycles and increased ecosystem resilience. In fact, forests contain 80 per cent of all the carbon stored in terrestrial vegetation, which in turn contains more carbon than in the entire atmosphere. For the time being, sustainable forestry in developing countries has emerged as one of the most cost-effective ways of mitigating climate change.

In terms of development benefits, forests play an essential role for poverty eradication, providing livelihoods for nearly 90 per cent of the 1.2 billion people living in extreme poverty worldwide. Developing countries produce about 25 per cent of the world's industrial wood products (i.e. sawn wood, panels, wood pulp and paper) and almost 90 per cent of its fuelwood.

Given this context, negotiations mandated under the UNFCCC Bali Action Plan are ongoing on the possible inclusion of 'Reduced Emissions from Deforestation and Forest Degradation' (REDD) in the next phase of a carbon trading regime, and specifically, the clean development mechanism (CDM), which smaller developing countries could benefit from directly. This would provide forest owners with a price for carbon storage and sequestration, which would be at least equal to or higher than what they get for selling timber, soya, palm oil, beef, sugar or maize on the international market. Meanwhile, populations could continue to benefit from sustainable use of the protected forests, deriving development benefits. REDD projects would also generate environmental benefits such as protection of ecosystem services, including soil protection, erosion control, water purification, reduced flooding, agricultural pollination, local rainfall and biodiversity protection, and human development benefits such as encouragement to resolve land tenure issues and increasing resilience and adaptation to climate change.

Carbon forestry – if fully credited under the CDM and the EU Emissions Trading Scheme (ETS) and integrated into an eventual international carbon market – could help to raise and effectively channel the capital needed for developing countries to take action to preserve their standing forests. A simple solution to the issue of 'avoided deforestation' at the international level would be to allow developing countries to be voluntarily listed in a new Annex under the successor Kyoto Protocol. These countries would follow current rules for land use and land cover-related emissions that exist between Annex I countries, while leaving the energy-related emissions for future consideration.

Capacity building needs for implementing REDD policies in smaller developing countries would need to be addressed, however, since they currently have limited capacity to monitor and account for changes in deforestation rates and emissions. Moreover, few have the capacity to develop participatory processes to ensure the engagement of relevant stakeholders. Technology transfer would also need to be stepped up in the forestry sector. A wide range of sustainable forestry management 'technologies', ranging from remote sensing assessment and monitoring technologies, integrated information management systems, harvesting and processing technologies exist. These technologies may be transferred in the form of knowledge, capital, goods and services, through, inter alia, trade, foreign direct investment, overseas development assistance and joint ventures.

Johannah Bernstein, Independent Consultant

Notes

1. Allison et al. (2005).
2. Workshop on Climate Change and Fisheries and Aquaculture, FAO Headquarters, Rome, 7–9 April 2008, Options for Decision Makers.

6

Competitiveness and Border Measures

6.1 Climate-Related Border Measures

Thomas L Brewer, Georgetown University, United States

Climate-related border measures to address industrial competitiveness concerns have been under discussion for several years. Border measures are highly controversial, as they are seen as 'sticks' rather than 'carrots' to encourage an inclusive approach to deal with the problem. Their legality under the World Trade Organization (WTO) has also been questioned. There are three dimensions to the border measures that are under consideration in the EU and US: the form the measures would take, the covered industries and the source countries covered.

With regard to the **form**, although the initial discussions in the EU several years ago focused on tariffs, the emphasis has since shifted to an emphasis on an allowance purchase requirement. The discussions in the US have focused on allowance purchase requirements for US importers. As to **industries,** only a small number of manufacturing industries are commonly identified in discussions in both the US and the EU – namely iron and steel, aluminium, cement, glass, pulp, paper, chemicals and ceramics – and therefore only the exports and industrial development of a small number of industries would be directly affected by any border measures that are ultimately adopted.

There has been a significant difference between US and EU discussions about the **countries** whose exports would potentially be covered. In the US, the concern has been with large developing countries that have not taken 'comparable actions' to mitigate greenhouse gas emissions. In contrast, in the EU the predominant concerns were initially focused on the US, because of its 'free-rider' status as a non-participant in the Kyoto Protocol. More recently, industry concerns have begun to include developing countries as well. The concern has been on a small number of large developing countries; there has not been any high-profile public discussion of exempting small developing countries from any possible future actions or of specifically including them.

There is uncertainty about the future 'border measure' policies of both the EU and the US. If and when such policies are officially adopted in the future, their relevance to individual developing countries will depend directly on the details of the policies and the basic facts of each individual developing country's economic situation and GHG emissions. Monitoring the specific relationship of each developing country to the details of EU and/or US policies will be essential. Least developed countries and countries with less than 0.50 per cent of world greenhouse gas emissions would be excluded from coverage under provisions in a prominent legislative proposal in the US. Therefore, it is likely that most developing countries will not be directly exposed to border measures on their exports.

Background

Interest in climate-related border measures has been increasing significantly since the Bali COP/MOP (Conference of the Parties, serving as a meeting of the Parties) climate conference in late 2007 and its Action Plan for the negotiation of a new multilateral agreement for the post-2012 period. The possibility of such measures had been under discussion for several years prior to that meeting in both the EU and the US, and they continue to be under active consideration.

In the EU, the issues have arisen from time to time during the past several years with regard to US non-participation in the Kyoto Protocol. The emphasis in the public discussions within the EU was initially on the possible imposition of offsetting tariffs, though the European Parliament's resolution (2005/2049) uses the generic term 'border adjustment measures'. The Commission's reaction to these measures was initially to oppose them on the grounds that they risked exacerbating trade relations with the US, particularly at a time when trade relations were already strained and when trans-Atlantic relations more generally were unusually conflicted over a broad range of issues. In addition, there have been concerns that such a measure would undermine support in the US among those political and business circles that have been hoping for increased EU–US co-operation on climate change issues. There have also been concerns that such measures might be challenged in a WTO dispute settlement case, and the outcome of such a case would inevitably be uncertain.

In November 2007 – in advance of the Bali climate change conference – the issue was again the subject of attention within the Commission and Parliament, and among industry and environmental groups. EU Enterprise Commissioner Günter Verheugen suggested that the Commission was more favourably inclined to address the issue through sectoral agreements, including perhaps voluntary global industry agreements – a position that has been supported by at least some industry and environmental organisations (see especially, *Financial Times*, 2007; and *EurActive*, 2007). Just before and after the release of the Commission's proposals for extension of the European Emissions Trading Scheme (ETS) on 23 January 2008, there was a resurgence of interest. Commission President José Manuel Barroso explicitly mentioned the possibility in a speech (European Commission, 2008a). The focus of discussion, however, has shifted away from tariffs to importers' purchases of emission credits.

At the same time, interest in border measures in the US has been increasing, particularly in the Senate in the aftermath of a favourable committee report on S.2191, America's Climate Security Act, introduced by Senators Lieberman and Warner. There has since been a 'substitute amendment' (S.3036) proposed by Senator Boxer, with the support of Senators Lieberman and Warner. This bill and other bills are likely to remain in a pending status until at least 2009 or 2010, when perhaps the new Congress and US President will enact legislation that will establish a US cap-and-trade system. Until then, S.3036 offers a useful glimpse into what future US legislation could entail. The bill includes provisions that would require US importers to purchase 'international reserve allowances' to offset low energy costs for imported 'covered goods' coming from 'covered countries' that have not taken adequate measures to mitigate greenhouse gas emissions. Its provisions are similar in many respects to those of another prominent climate change bill, S.1766 (commonly known as the Bingaman Specter bill), which requires reviews of whether countries have taken 'comparable action' to limit GHG emissions.

Types of border measures

There are three important dimensions to the climate-related border measures that are under consideration in the EU and US: the form the measures would take, the industries that would be covered and the source countries that would be covered.

As for the form, although the initial discussions in the EU several years ago focused on tariffs, the emphasis has since shifted to an emphasis on an allowance purchase requirement. The discussions in the US, as reflected in the proposed legislation in the Congress, have focused on allowance purchase requirements for US importers.

With regard to industries, Section 1311 of the Boxer substitute amendment (S.3036) to the Lieberman Warner 'Climate Security Act' lists the following key energy-intensive industries that would be covered: 'Iron, steel, steel mill products, aluminium, cement, glass, pulp, paper, chemicals, industrial ceramics'. In addition, however, it indicates that 'any other manufactured product sold in bulk that generates, in the course of manufacture, a substantial quantity of direct and/or indirect GHG emissions' could also be covered. Although there are no equivalent formalised lists of potentially covered industries in the context of the EU institutional policy-making process, the industries that are commonly mentioned in the EU are essentially the same as in the US.

As for the countries whose exports would potentially be covered, there has been a significant difference between the US and the EU discussions. In the US, the concern has clearly been with 'developing countries' – as that term is used in some versions of the proposed legislation. However, the names of individual countries are not mentioned. Rather, there are provisions for the imposition of border measures – if negotiations fail to resolve differences – on covered goods coming from countries that have not taken 'comparable actions' to mitigate greenhouse gas emissions. Methodologies and administrative procedures would be established within the executive branch of the government to determine the number of allowances that would have to be purchased by US importers in specific cases, based on a determination of the 'covered goods' and 'covered countries'.

In contrast, in the EU the predominant concerns were initially focused on the US, because of its 'free-rider' status as a non-participant in the Kyoto Protocol. Australia was also a concern in the EU, but is no longer since it became a participant in the Protocol in November 2007. The European Parliament resolution specifically referred to 'industrial' countries. More recently, however, industry concerns have begun to include developing countries as well, although such concerns have centred on a small number of large developing countries. There has not been any high-profile public discussion of excluding small developing countries from any possible future actions or of specifically including them.

Legal basis for border measures in the trade context

Whether any such climate-related border measures would become the subject of a WTO dispute case and what the outcome would be have received much attention, especially in the US in the context of deliberations about the legislation proposed in the US Senate. It is, of course, inherently difficult to make predictions about the prospects of such border measures in a WTO dispute process. However, it is possible to determine the key legal issues and identify features that the measures should and should not have in order to improve their chances of surviving a WTO dispute case.

The central issue is whether the measures would qualify as an allowed exception to non-discriminatory treatment under the terms of GATT Article XX. That article provides that 'Subject to the requirement that such measures are not applied in a manner which would constitute a means of arbitrary or unjustifiable discrimination between countries where the same conditions prevail, or a disguised restriction on international trade, nothing in this Agreement shall be construed to prevent the adoption or enforcement by any member of measures: ... (b) necessary to protect human, animal or plant life or health; ...'. The article also provides in section (g) that 'measures relating to the conservation of exhaustible natural resources' are exempted. Whether climate-related border measures would qualify as exceptions and other issues have been the subject of intense consideration by US Congressional staffs, lawyers, industry associations, think tanks, non-governmental organisations and scholars in recent years (see especially American Electric Power, 2007; and National Foreign Trade Council, 2007).

One of the design challenges for border measures stems from the fact that the driving force in the US for the inclusion of border measures in legislation establishing a mandatory cap and trade system is the domestic political economy concerns about the international competitiveness of a few industries. However, WTO rules would require that the intention and effectiveness of the measures be assessed in relation to environmental goals. In this respect, any prospective EU measures may be less vulnerable, because they were at least initially explicitly formulated to address the free-rider problem and thus concerned with the effectiveness of a multilateral environmental agreement. However, to the extent the EU concerns morph into a focus on international competitiveness issues, if any EU border measures are ultimately adopted, they may become more vulnerable to challenge in the WTO dispute settlement process.

Despite all the legal complexities and uncertainties, three features of the situation are apparent: first, there has already been a significant amount of legal analysis and nuanced development of legislative language in the US and the EU in order to minimise any border measure's vulnerability to a WTO challenge; second, there will always be residual uncertainty about what the ultimate fate of such a measure would be in the WTO; and third, there is widespread recognition that such a dispute in the WTO could do much damage to political support for both the climate regime and the trade regime.

Finally, it is important to note that a commonly expressed hope in the US and in the EU has been that border measures would not actually be used. Rather, they would merely be available in case they were perceived to be needed in specific circumstances that might evolve.

Legal basis for border measures in the climate context

The legal issues in the context of the multilateral climate have been less thoroughly analysed and are probably less problematic. Further, the existing provisions concerning trade in the key international climate agreements emphasise that trade policies should be 'open' and non-discriminatory, and they thus offer, if anything, grounds for questioning the use of border measures.

The most pertinent provisions are the following: Article 2.3 of the Kyoto Protocol notes that Parties should 'strive to implement policies and measures... in such a way as to minimise adverse effects, including the adverse effects ... on international trade...'. Article 3.5 of the UN Framework Convention on Climate Change (UNFCCC) notes that 'The Parties should co-operate to promote [an]... open international economic system' and that 'measures taken to combat climate change, including unilateral ones, should not constitute a means of arbitrary or unjustifiable discrimination or a disguised restriction on international trade.' Article 4.2 of the UNFCCC notes that 'measures taken to combat climate change, including international ones, should not constitute a means of arbitrary or unjustifiable discrimination or a disguised restriction on international trade'.

In sum, the key multilateral climate change agreements make clear that governments should generally adopt non-discriminatory trade policies and not use climate change concerns as a pretext for adopting protectionist trade policies. However, there are no enforcement mechanisms, and the statements are rather general.

In any case, the provisions are possible grounds for arguing that the climate agreements to date contain language that would tend to undermine the use of border measures.

Economic rationale

The economic rationale for establishing climate-related border measures is straightforward: countries that adopt measures to mitigate greenhouse gas emissions are imposing additional costs on industries, particularly those that are energy intensive. Although there can be significant variations in the costs among firms and individual installations in a particular industry within a country that imposes such measures, in general the firms and installations within the country face an additional cost that their rivals in countries that do not have GHG mitigating measures – or less stringent measures – do not face.

This generalisation is nonetheless subject to two important qualifications. First, the extent to which the international competitive issues arise from differences in countries' mitigation measures depends on the extent to which the goods are internationally traded, or tradable. Second, the cost differences would tend to diminish over time as installations in countries with relatively stringent mitigation measures adopt new technologies or otherwise change their production processes to make them less GHG-intensive. Thus, the economics of the international competitive issues vary across industries and over time.

Effectiveness of border measures

Assessing the effectiveness of border measures necessitates a specification of the objective. As noted, border measures can have one or more of three objectives: deterrence of free riders on multilateral agreements, reduction of leakage of GHG emissions and establishment of a level playing field among international competitors. Furthermore, the effectiveness will depend on the specific features of the measures.

Moreover, as far as competitiveness is concerned, one must be careful about the unit (or level) of analysis. Economists generally do not consider competitiveness to be a feature of national economies. In any case, while the competitiveness of one industry in a country may decline because of changes in the relative costs of electricity as a result of GHG-reducing measures undertaken in some countries but not others, the competitiveness of other industries will increase, because of the country's shifting comparative (dis)advantages as the relative prices of factors of production change.

Ultimately, the effectiveness of border measures in terms of their establishment of a level playing field among competitors will occur at the level of individual installations. The changes in the competitiveness of particular installations can occur within countries as well as among countries. For instance, the competitiveness of installations that are dependent on coal-fired power plants would be more vulnerable to GHG emissions mitigation measures than would installations that are dependent on nuclear or hydro plants – whether they are located in the same country or a different country.

Implications for developing countries' exports and industrial development

The key determinants of how climate-related border measures would affect developing countries' exports and industrial development are first, which specific products are included in the covered industries, and second, how extensively the greenhouse gas emissions are traced through stages in the production process.

Since the administrative criteria and procedures for determining these issues are not yet known, we can only speculate at this point. However, it seems likely that if there are eventually any border measures in place, they will cover the list of industries noted above – namely iron and steel, aluminium, cement, glass, pulp, paper, chemicals and ceramics. Exactly how those industries and their specific products will be defined operationally remains to be seen. Presumably, they would be identified in terms of their Harmonized System (HS) codes. In any case, the discussions thus far have been only about goods and not services.

The extent to which the various phases of production processes will be included in the computations of greenhouse gas emissions poses more complex issues. In particular, will the (indirect) greenhouse gas emissions, including those of multiple suppliers in multiple countries, be included, in addition to the direct emissions in the final production stage in the exporting country? Imagine, for instance, aluminium being produced in a country that has both hydro-power plants and coal-fired power plants. Clearly, the greenhouse gas emissions will depend which of these is the source of the electricity for aluminium production at a particular facility. Further, the bauxite could come from another country, so that the greenhouse gas emissions emitted in the mining and transportation of the bauxite to the aluminium production facility could also be included.

The exposure of developing countries' exports and industrial development, therefore, will depend not only on the industry structure of their economies and exports, but also on the industry lists and GHG calculation procedures used in the importing countries.

The exposure of developing countries will also depend on the nature and extent of their own measures to mitigate greenhouse gases. If they were considered to have taken 'comparable action' to the US, then of course there would be no border measure imposed on their exports to the US. At the same time, two categories of countries are excluded from coverage: 'least developed' countries (according to the UN definition) and countries with de minimis emissions (i.e. those with less than 0.5 per cent of world emissions). See tables 6.1 and 6.2 for lists of the 49 and 129 countries, respectively, in those categories.

As for the future, of course, the applicable criteria and the lists of excluded countries can change. In that context, it will be important to keep track of both (i) changes that occur in the legislation or implementing administrative procedures in the United States, and (ii) changes that occur in the developing countries themselves, including where they remain on official lists of least developed countries and whether the relative levels of their GHG emissions change. Further, it should be noted that changes in the US classification of countries for trade purposes; for instance, the 'graduation' of a country from the Generalized System of Preferences (GSP) (see World Resources Institute, 2008) does not necessarily imply changes in a country's position relative to the lists of countries that are excluded depending on whether it is on the official UN list of least developed countries, or perhaps more importantly whether it is excluded on the grounds of its relatively small GHG emission levels. Again, however, the application of

US policy on a particular developing country will be determined on the basis of the precise language of any legislation that is eventually adopted by the US – which is still probably at least a year or two in the future – and the precise administrative procedures in place, in combination with the exact situation of any individual developing country. It will thus be important for all countries to monitor future developments – even including those that are small, least developed, islands and/or otherwise especially vulnerable to climate change.

Table 6.1 UN list of least developed countries

Afghanistan	The Gambia	Niger
Angola	Guinea	Rwanda
Bangladesh	Guinea-Bissau	Samoa
Benin	Haiti	São Tomé and Principe
Bhutan	Kiribati	Senegal
Burkina Faso	Lao People's Democratic Republic	Sierra Leone
Burundi	Lesotho	Solomon Islands
Cambodia	Liberia	Somalia
Central African Republic	Madagascar	Sudan
Chad	Malawi	Timor-Lesté
Comoros	Maldives	Togo
Democratic Republic of the Congo	Mali	Tuvalu
Djibouti	Mauritania	Uganda
Equatorial Guinea	Mozambique	United Republic of Tanzania
Eritrea	Myanmar	Vanuatu
Ethiopia	Nepal	Yemen
		Zambia

Note: If US legislation entered into force in the future with an exclusion for climate-related border measures based on a UN list, the countries on the list could obviously differ.

Source: UN List of Least Developed Countries, obtained from www.un.org on 14 June 2008.

Table 6.2 Countries with less than 0.50 per cent of world GHG emissions (2000)

Egypt	Sudan	Angola
Colombia	United Arab Emirates	Israel
Kazakhstan	Korea (North)	Peru
Malaysia	Iraq	United Republic of Tanzania
Philippines	Myanmar	Kuwait
Vietnam	Dem. Rep. of Congo	Syria
Algeria	Belarus	Ethiopia
Bangladesh	Chile	Turkmenistan

Bolivia	Jordan	Sierra Leone
Morocco	Lebanon	Eritrea
Serbia and Montenegro	Bosnia and Herzegovina	Guyana
Libya	Guinea	Mauritius
Singapore	Cambodia	Suriname
Azerbaijan	Laos	Swaziland
Kenya	Afghanistan	Liberia
Cuba	Bahrain	Burundi
Uruguay	Mauritania	Lesotho
Ecuador	Honduras	Fiji
Nepal	Nicaragua	Malta
Madagascar	Jamaica	Equatorial Guinea
Cameroon	Macedonia, FYR	Guinea-Bissau
Côte d'Ivoire	Niger	Belize
Paraguay	Botswana	The Bahamas
Qatar	Costa Rica	Antigua and Barbuda
Zimbabwe	El Salvador	Djibouti
Oman	Brunei	Barbados
Tunisia	Congo	The Gambia
Dominican Republic	Moldova	Bhutan
Yemen	Panama	Seychelles
Trinidad and Tobago	Benin	Maldives
Zambia	Tajikistan	Vanuatu
Mongolia	Namibia	Cape Verde
Uganda	Georgia	Saint Lucia
Guatemala	Haiti	Samoa
Croatia	Papua New Guinea	Comoros
Mali	Kyrgyzstan	Solomon Islands
Central African Republic	Cyprus	Grenada
Sri Lanka	Togo	Palau
Ghana	Malawi	St Vincent and the Grenadines
Burkina Faso	Gabon	Tonga
Chad	Albania	Dominica
Mozambique	Armenia	St Kitts and Nevis
Senegal	Rwanda	São Tomé and Príncipe

Note: The countries are listed in descending order column by column according to the percentage of world GHG emissions, starting with Egypt, which was 0.49 per cent in 2000. The list is based on emissions data for 2000. If US legislation entered into force in the future with an exclusion to border measures, the countries on the list could obviously differ. Also note that although the source is widely used and generally considered a reliable source, it is from an environmental organisation, not an official governmental or international agency. The list therefore does not necessarily correspond to US or other formal lists of countries.

Source: World Resources Institute, Climate Analysis Indicators Tool, obtained from www.wri.org on 14 June 2008.

Issues for developing countries to consider

Which approaches could be pursued within groupings of developing countries, including African, Caribbean and Pacific countries, least developed countries, small vulnerable economies and small island developing states, to achieve optimal policy outcomes with respect to these issues at the interface between climate change and trade?

The first step is to recognise that any climate-related border measures – if and when they are eventually adopted – may not have direct effects on many, or even most, developing countries. This is because the focus of attention in the EU has been on the US as a free rider on a multilateral agreement, and because least developed countries and countries with less than 0.50 of world GHG emissions would be excluded from such measures on the basis of the proposed legislation in the US Senate that has become the furthest advanced in the legislative process. Therefore, it is likely that most developing countries will not be directly exposed to border measures on their exports. Of course, only the precise provisions of actions concerning border measures – if any – will ultimately determine what countries are and are not excluded from coverage. In any case, developing countries could consider the possibility of supporting the inclusion of border measures in a post-2012 multilateral climate regime. Such provisions could include exclusions for least developed countries and countries with very low levels of GHG emissions.

It is important to distinguish among different, though related and overlapping, objectives that can motivate the use of border measures in the context of climate change issues. They can be formulated to address free-rider problems in multilateral climate agreements, and/or greenhouse gas leakage problems in specific combinations of countries and/or international competitiveness issues in specific industries.

There have been significant differences in the discussions and the extent of the development of the details of any proposed border measure policies in the US and the EU. However, in neither case has the discussions yet resulted in legislation that would impose border measures. Even if border measures were eventually included in formally stated policy in the US and/or the EU, this would not necessarily mean that such measures would actually be implemented in practice. For the implementation would depend on the details of the administrative procedures and the factual circumstances of individual cases.

Only a small number of manufacturing industries are commonly identified in discussions on border measures in both the US and the EU, and therefore only the exports and industrial development of a small number of industries would be directly affected by any such measures should they ultimately be adopted. Exports of covered goods from covered countries could nevertheless occur, even if they were subject to border measures, if their production processes were sufficiently efficient and not overly greenhouse gas intensive, so that they could still be price-competitive relative to domestically produced goods in the US or EU.

Least developed countries and countries with less than 0.50 per cent of world greenhouse gas emissions would be excluded from coverage under provisions in a prominent legislative proposal in the US. Whether these exclusions - or similar ones - will be included in any legislation that is ultimately passed in the US or in any measures taken that might be taken in the EU, of course remains to be seen.

As noted above in regard to several specific issues, there is uncertainty about the future 'border measure' policies of both the EU and the US. If and when such policies are officially adopted in the future, their relevance to individual developing countries will depend directly on the details of the policies and the basic facts of each individual developing country's economic situation and GHG emissions. Monitoring the specific relationship of each developing country to the details of EU and/or US policies will be essential.

Climate-related border measures are all obviously high-stakes issues, with potentially significant consequences for many developing countries. They are also likely to become increasingly salient issues in both international climate forums and international trade forums for at least the next several years, and perhaps for the indefinite future. They are therefore issues on which representatives of government, business and civil society will all need to become more knowledgeable.

6.2 Energy-Intensive Sectors and International Trade: Possible Sectoral Agreements

Joyashree Roy, Jadavpur University, India and Samantha Fang, ICTSD

International competitiveness concerns have led to renewed interest in sectoral agreements covering energy-intensive and heavily trade-exposed industries such as the steel, pulp and paper, aluminium, cement and agrochemical industries. Though debated, sectoral proposals recognise the need to broaden the scope of GHG mitigation and engage developing countries that are reluctant to accept mandatory emissions cuts. Proponents argue that sectoral approaches should be considered as part of a larger international framework for climate change mitigation, with the goal not only to alleviate competitiveness concerns but also to facilitate the diffusion of best-available technology.

The focus on large emitters in current discussions about sectoral approaches - with particular emphasis placed on the five large developing countries Brazil, China, India, Mexico and South Africa - has taken attention away from the concerns of small developing countries (SDCs). The urgent need for a successful climate change framework that includes the large developing countries, and the recognition that mitigation efforts are most efficient when tailored for these states, have caused the mitigation potential and future concerns of SDCs to be neglected. Because of their traditionally low emissions, SDCs do not face mandatory emissions cuts under the Kyoto Protocol, and to date they have not been actively engaged in the 'universal' sectoral approaches negotiating process.

Sectoral approaches can provide a means to engage major emitters by incentivising mitigation efforts with mechanisms for technology transfer and financing. Sectoral approaches arguably could become a tool for the development and deployment of efficient technologies, which are considered to be a key component in future global agreements. Benchmarks set by 'best practice' and 'best available technology' can drive the most efficient technology to developing countries in specific sectors. Benchmarks, set by governments and met by businesses, allow for public–private partnerships that can identify and implement best practices in a joint learning process.

Given high levels of natural minerals export in small developing countries, it is strategic in the long run for developing countries to adopt mitigation efforts, and adopting sectoral emissions targets would help in harmonisation of technology deployment while helping to meet growth aspirations. Clean development mechanisms (CDMs), as negotiated under sectoral approaches, may in fact offer a level of flexibility that can allow for sustainable development needs, lowering the transaction costs of CDM through the bundling of small-scale projects in an individual sector. However, it is important to recognise that sectoral approaches cannot eradicate the intrinsic flaws of CDMs, i.e. lack of domestic participation and an incentive structure based on climate benefits but not other sustainability benefits.

Given that many are resource-rich, small developing countries could view emissions mitigation under sectoral approaches as an opportunity to upgrade export capacity from raw mineral products to finished goods. Participation in a sectoral approach regime could mark the entry of resource-rich small developing countries into energy-intensive industries. The use of 'best-available technologies' through the sectoral initiative conforms domestic production to international standards, essentially linking an individual market to the global value chain. Participation could be helpful to small developing countries if they use backward linkages and capacity building in keeping with broader, sustainable development goals.

Though not a new concept in climate negotiation literature, sectoral approaches (SAs) have gained new prominence with the 2007 Intergovernmental Panel on Climate Change (IPCC) assessment of sectoral mitigation potential and their inclusion in the 2007 Bali Convention. While there are conflicting interpretations as to what SAs might entail, SA proposals focus on transnational, across-the-board mitigation of energy-intensive industries. The discussion on SAs stems from international competitiveness concerns, and proponents argue that SAs can enhance mitigation efforts by making participation broad based, facilitating technology transfer that can promote best-practice technologies and preventing emissions leakage.

This section reorients the discussion and focuses on the concerns of small developing countries (SDCs), which include the least developed countries (LDCs), small vulnerable economies (SVEs) and small island developing states (SIDS). Given the principle of 'common but differentiated responsibilities', it is clear that SDCs do not face mandatory emissions cuts in the Bali Convention. However, SDC concerns should not be overlooked in SA negotiations, and there is possibility that expanding the scope of

SAs for broad-based participation can facilitate better and more informed discussion on a post-Kyoto framework.

As such, SDCs can play an integral role in SA negotiations, ensuring that SAs reflect their interests and long-term development goals. In fact, SDCs can have great commonality of interest with developed countries in adopting sectoral approaches that can serve as a springboard for further international consensus. With voluntary participation, they can realise the potential gains of sectoral approaches, which can facilitate technology transfer, utilise clean development mechanisms (CDMs) for broader development goals and enhance capacity building.

Sectoral approaches

International competitiveness concerns have led to renewed interest in sectoral agreements (SAs), covering energy-intensive and heavily trade-exposed industries such as the steel, pulp and paper, aluminium, cement and agrochemical industries. Though debated, SA proposals recognise the need to broaden the scope of GHG mitigation and engage developing countries that are reluctant to accept mandatory emissions cuts. Bali Action Plan 1(b)(4) has opened discussion on SAs as a part of a larger international framework for climate change mitigation, with the goal not only to alleviate competitiveness concerns, but also to facilitate the diffusion of best-available technology. However, even proponents of SAs caution that it is an imperfect solution, one that should be discussed as a complementary element to other reduction mechanisms in a framework that balances mitigation and adaptation. Japan introduced a sectoral proposal in February 2008 and made this issue central in the G8 Hokkaido Toyako Summit in July 2008, though no substantive conclusions were reached.

Types of sectoral approaches

Various stakeholders have varying definitions as to what sectoral approaches constitute and how they might function, but three major proposals have emerged from the discussion. First, sector-wide transnational approaches entail the negotiation of multiple sector agreements that, when taken collectively, cover a significant share of total emissions. Industry actors in the given sector would shape the form and the substance of these agreements, and industry-led initiatives could push for global industry standards (Stigson et al., 2008).

Discussion has also begun on 'bottom-up' country commitments, where emissions targets are set for individual sectors using best-available technologies (BATs) and then added for a quantified national target. Honouring the 'common but differentiated responsibilities' of countries under the UNFCCC, policy-makers have discussed sector emissions cuts only for developed countries; however, SAs are expected to engage additional countries currently not covered by the Kyoto Protocol. Because targets are based on a sector-specific assessment (with parameters such as output capacity and feasible sectoral technology), the bottom-up approach lends confidence to achieving credible targets.

Critics of the proposed 'bottom-up' approach state that it would be ineffective because it is too lenient on industries. Yet the Japanese government has argued that sector-specific targets would make it easier and more viable for China, India and United States to participate. Japan has proposed setting mid-term national targets for 'each major emitting country', but calculating and amounting emissions reduction potential in sectors such as power-generation, transport and other energy-intensive industries using certain indicators.

A third model delineated is the 'top-down' sectoral crediting model, offering incentives such as a sectoral clean development mechanism (CDM) as incentives for emissions cuts (Stigson et al., 2008). The voluntary 'no lose' pledge would allow developing countries to achieve a certain GHG intensity target in carbon-intensive industries. Emissions achieved beyond voluntary targets would be eligible for sale as emission reduction credits to developed countries, while failure to meet the voluntary pledge level would not involve any penalties. Furthermore, in order to encourage developing countries to pursue more aggressive mitigation targets, developed countries and international financial institutions could pledge to provide finance and technical assistance.

However, recent literature has shown that these three models are imperfect, and stakeholder engagement may in fact compromise the competitiveness and reduction goals of SAs. Furthermore, sectors themselves are difficult to define: governments and organisations such as the IPCC, IEA and UNFCCC define sectoral boundaries according to their own priorities for data collection, analysis and discussion (ICC, 2007). Complex product value chains also add to the difficulty of sector identification and delineation.

Potential benefits of sectoral approaches

Proponents of SAs argue that they offer many benefits aligned with developing country goals and can contribute to post-2012 efforts by increasing transparency, defusing competitiveness issues, engaging broader participation (most importantly from businesses in emerging economies) and promoting the diffusion of best practices and technology. The SA is designed to reduce carbon leakage by co-ordinating action and policies across countries according to an overall plan and target for specific industries. Such approaches can identify emissions on a sector-by-sector basis: policy-makers then can identify abatement potentials and set goals for specific industries, rather than adopting blanket goals for each country. Some contend this method is more equitable and attainable, building confidence and leading to political acceptability by national governments.

In general, the information data collected in the SA process can be beneficial in identifying BATs and facilitating their dissemination, as well as providing concrete baselines for mitigation. Performance indicators expressed through industry benchmarks can discover unknown abatement potentials in developed and emerging economies to realise more cost-effective solutions. This process allows policy-makers to target efforts and address first the sectors where action is more urgent and international co-operation is most critical. However, key to the success of an SA and for adequate transparency that can allay competitiveness concerns is the presence of effective monitoring, reporting and verification (MRV) systems. Developing countries can find that effective systems are costly and time consuming, and as the EU ETS experience shows, they require significant industry and government capacity, which many crucial emerging economies lack.

Keeping in mind the idea of 'common but differentiated responsibility', SAs can provide a means to engage major emitters while incentivising and compensating mitigation efforts with mechanisms for technology transfer and financing. SAs arguably could become a tool for the development and deployment of efficient technologies, which is considered to be a key component in future global agreements. Supporters point to the, albeit limited, technology-sharing currently facilitated by the International Iron and Steel Institute (IISI) as an example of the potential for SAs. Theoretically, benchmarks set by 'best practice' and defined by the 'best-available technology' (BAT) can drive the diffusion of existing technology to developing countries in specific sectors. At the same time, government–industry partnerships can work to identify and implement best practices in a joint learning process that allows for further collaboration in emissions mitigation. SAs can be used to identify 'win-win' opportunities to improve technological and operational efficiency, facilitating co-operation between companies in a sector to develop breakthrough technologies in proprietary industries.

SA proposals should note the importance of assisting developing countries in achieving domestic policy objectives, co-operating with these countries in a way that suits their interests. Possible incentives for developing countries might include sectoral crediting and no-lose targets, as outlined in the 'top-down' approach, as well as capacity building. However, sectoral crediting requires that the supply of credits needs to be matched by demand, resulting in more unilateral commitments from developed countries. Also, because SAs are sector-wide co-ordinated activity, they also have potential antitrust implications for national and regional jurisdictions. Sectoral crediting could potentially undermine competition in developing countries if select companies are awarded credits, creating barriers to entry and worsening the investment climate. Capacity building – which theoretically provides access to data and best practice, technology co-operation and transfer and technical assistance – is still the most compelling incentive for a sectoral approach.

Continuing concerns at the international level

General concerns about SAs, as well as specific developing country concerns, hamper the acceptance and adoption of SAs. First, businesses lack incentives to collaborate with potential competitors in the same sector. Governments still need to find adequate rationale for inducing individual companies in the industry to reveal information. In designing GHG reductions, the asymmetry of information between government and business may hamper efforts toward such reductions, as business possibly understates the industry's ability to adjust processes and invest in new technologies. The International Energy Agency's (IEA) work on energy efficiency potentials shows lack of sound international data for individual sectors, leading to the possibility of gaming as governments have limited knowledge of technical details or heavy industrial activities. Further, business can lobby governments for special treatment according to particular circumstances, possibly creating CO_2 havens that are counterintuitive to the approach. Therefore, in order for sectoral approaches to be effective in information collecting, technology diffusion and ultimately emissions reduction, governments must incentivise business to be a reliable partner in the process.

More relevant to developing countries, the political negotiating realities of the various stakeholders under the UNFCCC framework make it difficult to reach a consensus about SAs. SAs can address concerns about competitiveness and carbon leakage only if benchmarks are set across all countries – including developing countries – in the specific industry. It is becoming increasingly clear that GHG goals require contributions from developing countries, and unilateral action by Annex I countries cannot ensure stabilisation (Schmidt et al., 2006). Furthermore, The World Energy Outlook shows that the potential for energy efficiency improvements until 2030 are greater in developing countries than in OECD countries, necessitating a means to engage these regions in measures and activities that would reduce emissions and increase energy efficiency (Baron, 2006).

However, developing countries are wary that sectoral benchmarks might open the backdoor to new standards and obligations, which could hinder or complicate developing country exports. They argue that developed countries – and especially non-participating Annex I major emitters such as the United States – need to shoulder commitments and that further obligations do not fall under the agreed 'measurable, reportable and verifiable actions' to be undertaken by developing countries.

As such, SAs have not been widely endorsed in climate change negotiations, though discussion on linking SAs to a larger climate change framework has begun with papers by the Pew Centre and the Centre for Clean Air Policy (CCAP). What remains is for developing countries to negotiate terms about SAs that will suit their interest, and for the international community to find synergies between emissions mitigation for all countries and broader, sustainable development goals.

Current players and concerns

Because of their low production of manufactured goods, and in many cases the fact that they are not present in energy-intensive industries, small developing countries have in large part been marginalised from sectoral approaches discussions. Indeed, SDCs arguably represent an insignificant share of world trade, making up only 4.25 per cent of world merchandise exports (see table 6.3, below). They therefore have very little impact on the international trade regime through their exports and imports.

In presenting the concerns of SDCs, it is important to recognise that there are great differences between LDCs, SVEs and SIDS, and one must take caution before prescribing generalised recommendations for the entire category. For example, the least developed countries group is extremely diverse. In a group of LDCs surveyed published by the International Institute for Environment and Development (IIED), 15 LDCs had an annual growth rate of real GDP per capita of more than 2 per cent; yet another group of 22 LDCS has been stagnating or regressing since the 1990s (Huq et al., 2003). However, despite their differences, the category of small developing countries shares common characteristics: their economies are highly vulnerable to external shocks and natural disasters, and many such countries are critically dependent on external finance. For many SDCs, development prospects are based on aid relationships and external debt dynamics, and fundamental issues of capacity building remain unaddressed. Grouping SDCs, despite their differences, therefore serves to distinguish them from the large developing countries, whose concerns have thus far dominated climate change discussions.

However, access to world markets for exports is critical to the economies of many of these countries, and any discussion on a regime that will affect particular industries and adopt standard targets across sectors – as proposed by discussion about sectoral approaches – will have a great impact on SDCs. It is therefore necessary to voice SDC concerns as distinct from those of larger developing countries, rather than allow their issues be subsumed into broader climate change framework discussions.

In absolute terms, carbon emissions rose for SDCs (including Hong Kong and South Korea) in the 2000–2005 period from 4925.97 million metric tonnes in 2000 to 6031.28 million metric tonnes in 2005. Discounting Hong Kong and South Korea, SDC emissions in absolute terms rose from 4430.62 million metric tonnes in 2000 to 5456.59 million metric tonnes in 2005. However, their relative percentage of world emissions has not changed, remaining constant at 19 per cent. What is interesting is that the SDC growth rate, as measured by a higher share in global GDP, has increased relative to more mature economies such as Hong Kong and South Korea with lower emissions. This fact may be explained either by convergence or, more likely, by the structural bias of these economies toward the production of manufactured goods (see figure 6.1).

While SDCs have contributed the least to greenhouse gas emissions, they remain the most vulnerable countries to climate change and have the least capacity to adapt to fluctuations in environment. The IPCC provides the following necessities for adaptation: a stable and prosperous economy, a high degree of access to technology at all levels, well-delineated roles and responsibilities for implementation of adaptation strategies, systems in place for the national, regional and local dissemination of climate change and adaptation information, and an equitable distribution of access to resources. As at the time of writing, SDCs lack the necessary institutional, economic and financial capacity to cope with climate change and to rebuild infrastructure after natural disasters. IIED argues that SDCs, and in particular LDCs, are dependent on external

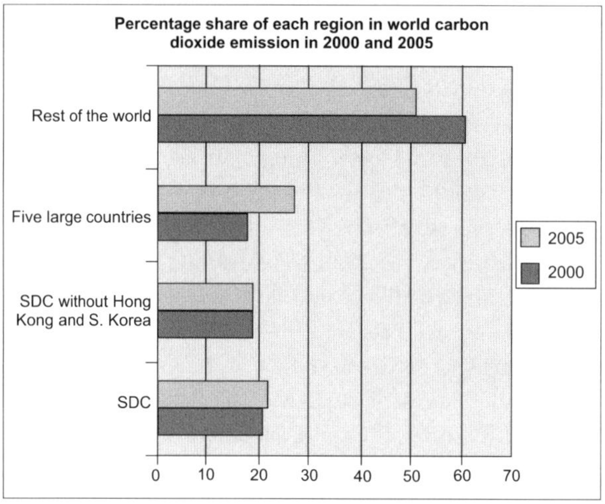

Percentage share of each region in world CO_2 emission		
	2000	*2005*
SDCs	21	22
SDCs without Hong Kong and S. Korea	19	19
Five large countries	18	27
Rest of world	61	51

Figure 6.1 Regional CO_2 emissions in 2000 and 2005

Source: http://www.eia.doe.gov/pub/international/iealf/tableh1co2.xls (Authors' estimates) [last accessed 13 March 2009].

Table 6.3 Regional percentage shares for GDP, population, and exports and imports

Year 2006	*Percentage share of each region in world GDP*	*Percentage share of each region in world population*	*Percentage share of each region in world merchandise exports*	*Percentage share of each region in world merchan dise imports*
LDC	0.69	11.51	0.85	0.79
SIDS and SVEs	1.15	3.82	3.40	3.43
Five large countries	11.80	42.25	12.69	11.33
Rest of the world	86.36	42.41	83.07	84.45

Source: Authors' estimates.

Table 6.4 Regional data for GDP, population, and exports and imports

Year 2006	*GDP in current US$ (million)*	*Population in million*	*Merchandise exports in US$ (million)*	*Merchandise imports in US$ (million)*
LDC	334,598	752.828	102,859	98,688
SIDS and SVEs	557,300	250.002	411,078	425,896
Five large countries	5,718,400	2,762.5	1,536,220	1,408,005
Rest of the world	41,851,602	2,772.77	10,057,843	10,494,411
World	48,461,900	6,538.1	12,108,000	12,427,000

Source: Authors' estimates.

aid, lack financial resources to absorb environmental shocks and cannot implement prevention and adaptation studies because of insufficient funds.

Political negotiating realities

Analysis of the SDCs as a group shows that while they are dominated by oil-importing countries, they also have an almost equal share with the rest of the world in global fuel and mining exports. The importance of their natural mineral export to the rest of the world, which is necessary to manufacture the energy-intensive goods targeted by proposed sectoral approaches, has implications for the impact of sectoral approaches on SDCs. First, the amount of natural mineral export attests to the abundant natural resources of these countries. The disparity between oil use (as characterised by import) and mineral export, as well as between manufactured good export and mineral export, shows that SDCs are by-and-large producing non-value-added goods and relying on raw materials for foreign exchange. For example, while SDCs currently export 45 per cent of the world's fuel and mining material, they only contribute 1.27 per cent of the world's manufactured iron and steel, 0.84 per cent of pulp and paper and 2.79 per cent of aluminium.

These two realities have several possible conclusions. They spell the possibility of SDC market entrance in the energy-intensive industries discussed under sectoral approaches, as increasing transport costs incentivise the relocation of finished product manufacturing closer to raw material sources. Second, given rising population growth and energy demand, it is possible that SDCs can leverage their natural mineral exporting strength in future international discussions. SDCs therefore have a large role to play in future energy-intensive industry discussions, and it is important to include them in current discussions about sectoral approaches and to engage them as partners in emissions mitigation.

Given their high levels of natural mineral export, we project that any increase in oil prices will put SDCs in a better position compared to other larger developing countries. Furthermore, inflation would be higher in larger developing countries, leading us to believe that these large developing countries – which sectoral approach discussions are mainly targeting at this point – have less incentive to adopt a carbon-constrained climate change regime. Because of rising oil prices, dependence on many SDCs for mineral exports and inflation concerns, we hypothesise that there is high confidence of short- and long-run competitiveness loss for large developing countries relative to SDCs.

As such, there is strategic interest for small developing countries to adopt mitigation targets, and they may in fact have greater commonality of interest in emissions cuts than the larger developing countries that now play an important role in climate change negotiations. Our projections show that a long-term global carbon constraint and burden-sharing principle (with a span until 2050) needs to be defined not only for developed countries, but also for large developing countries and SDCs. Given the growth aspirations of developing countries as a bloc, there will continue to be barriers from large developing countries to participate and accept carbon constraints. However, for SDCs, the expected gain in competitiveness because of high natural mineral export may make them more conducive to mitigation efforts.

We must remember though that SDCs are major importers of manufactured goods, and unless there is longer-term investment in the domestic economy, a decline in competitiveness in that sector may actually trigger non-participation in the short run. As a result, countrywide targets may not be acceptable for SDCs, though they are likely to place more importance than large developing countries such as India and China on adopting sectoral emissions targets; these in turn would help in the harmonisation of technology deployment and help them meet their growth aspirations.

SDCs as important players

At the time of writing, Annex I countries clearly dominate the list by a large margin in both the export and import of energy-intensive goods. However, comparing the five large developing countries with SDCs (table 6.5) yields that SDCs are ahead of large developing countries, with a higher share in the export and import of products currently under discussion in sectoral approaches negotiations.

In the period of 2002–2006, SDCs were larger players than the large developing countries in the global export and import of products such as pulp and paper and aluminium, and their trade in these products continues to rise. The trend was reversed only in the case of iron and steel exports, in which larger developing countries increased their share of export from 9 per cent to 15 per cent, while SDCs' share declined from 19 per cent to 18 per cent. This trend in energy and steel was largely driven by the construction boom in China and India.

Table 6.5 Countries that are competitive in energy-intensive sectors

Exports	*Share in world exports of iron and steel (%)*	*Share in world exports of paper and pulp (%)*	*Share in world exports of aluminium (%)*	*Share in world exports of fertilisers (%)*
SDCs (2002)	19	13	16	26
SDCs (2006)	18	14	17	27
SDCs without Hong Kong and South Korea (2002)	14	9	13	25
SDCs without Hong Kong and South Korea (2006)	13	10	14	26
Five large countries (2002)	9	4	8	4
Five large countries (2006)	15	6	12	6
Rest of the world (2002)	72	83	76	70
Rest of the world (2006)	67	80	71	67

Given this data, it seems that the excessive focus on the large developing countries is misplaced and may not be efficient for reaching consensus in international negotiations. While the shares of the large developing countries are increasing in energy-intensive goods, they are doing so at a lower rate when compared to SDCs. Therefore, since Annex I countries dominate the global market for energy-intensive goods, it is clear that all developing countries should be given preferential treatment in catching up, but with emphasis placed on the SDCs that are leading the group.

Opportunities and challenges

Because of their limited emissions, low market shares and development status, SDCs cannot be politically bound to cut emissions under any mitigation regime. As a result, existing literature about mitigation efforts has focused on incentives for developing countries, searching for areas of possible consensus and synergies where both emissions mitigation and sustainable development can be expedited. SDCs could therefore participate in a sectoral approaches regime on a voluntary basis, benefiting from its associated technology transfer, improved CDMs, backward linkages and capacity building, which could fit within each SDC's broader sustainable development goals.

Technology transfer

Of the many foreseen benefits of sectoral approaches, the diffusion of efficient technologies is among the most important to the growth of developing countries. Opportunities for non-Annex I countries depend integrally on the speed at which technology transfer can happen, as well as the ability for recipient countries to absorb and integrate the knowledge that comes with technology. As described before, because SAs set targets based on 'best practice' and best-available technologies (BATs), participation in a sectoral approaches regime would theoretically drive the adoption of the most up-to-date technologies in a targeted sector.

Producers may in fact see an opportunity in increasing efficiency through the deployment of efficient technologies if they can become competitive with imports. Because SDCs are largely net importers in the current regime, this fact may turn the terms of trade in favour of them. As a result, efficient producers may have an incentive to look for new destinations for their investments, facilitating technology transfer, creating good opportunities for SDCs with low capacity and current low rates of technology transfer.

Furthermore, it can be expected under sectoral approaches that the carbon constraint on energy-intensive, finished products from developed countries would effectively increase their prices. Because most SDCs are net importers of finished products, this trend could act as an incentive to reduce imports and focus more on domestic production. The trend toward domestic infrastructure building, as well as incentivised technology diffusion under sectoral approaches, could be a good opportunity for SDCs to receive the technology necessary for capacity building and technological catch-up.

Clean development mechanism

The clean development mechanism (CDM), a project-based mechanism, was codified under the Kyoto Protocol to assist non-Annex I countries in achieving sustainable development, giving Annex I countries emissions credits (Certified Emissions Reductions, or CERs) toward their Kyoto targets in exchange for mitigation projects in non-Annex I countries. While CDMs have recently come under intense criticism for their complexity and high cost, CDMs continue to offer a concrete way for the diffusion of energy-efficient technology and processes from developed to developing countries.

However, because of their incentive structure, CDMs suffer from geographic imbalance, with projects so far concentrated in a limited number of countries (Sterk and Wittneben, 2006). In 2006, 411 (almost 36 per cent out of 1,145 projects) were hosted in India. Asia hosted 687 projects, Latin America had 415, while sub-Saharan Africa, the location of the vast majority of least developed countries, hosted only 17 projects (Sterk and Wittneben, 2006). This imbalance shows a skewed preference toward major emitters, while effectively discounting small developing countries, which could benefit greatly from these projects.

The main problem with CDMs under the current regime is that developed countries are incentivised to invest in projects that yield CERs rather than meaningful benefits for the host country. As of the time of writing, CDM projects have limited transformational effects, as they do not have a formal way to support host countries in formulating, enforcing and monitoring sustainable development (Figueres, 2006; Sterk and Wittneben, 2006). As a result, it is difficult for local stakeholders to be meaningful involved or for the project to contribute to capacity building and the host country's broader development goals. In some cases, projects may not even necessarily meet the conditions of the developing countries, perhaps disregarding or even violating their needs (Sterk and Wittneben, 2006). Because CDMs only reward the project's climate benefits, CDM projects are developed only with generated CERs in mind; this benefits the developed country that initiates the project, and thus does not have the primary goal of sustainable development.

Nor have CDMs fully fulfilled the goal of technology transfer, while they have given rise to other economic development distortions (Sterk and Wittneben, 2006): experiences under CDMs have shown that high transaction costs and knowledge intensity make the mechanism an artificially crafted market mechanism, making it difficult for all developing countries to be beneficiaries.

In this light, CDMs as negotiated under sectoral approaches may in fact offer a level of flexibility that can allow a developing country to formulate projects consistent with its sustainable development needs. For example, SAs could facilitate local policy initiatives in a sector: a host government could commission local companies on an emissions reduction project in accordance with its development goals. Again, because developing countries are not subject to mitigation cuts under the UNFCCC, CER credits could flow directly to the host government, which could redistribute the credits to partner firms and countries. Furthermore, because projects are locally selected, it is possible for such CDM projects to contribute to capacity building because they are chosen to have an impact on local environmental and social conditions and therefore have possibilities for meaningful involvement by local stakeholders.

Sectoral approaches can further contribute to CDM flexibility by facilitating the bundling of related projects in a sector. Through bundling, several small-scale CDM project activities are grouped together into a single CDM project activity, without the loss of distinctive characteristics of each activity. Bundling projects under the sectoral approach to the CDM would allow the CDM to further its development mandate, achieving not only local changes, but also 'sector-wide transformations' in host countries (Sterk and Wittneben, 2006). Developing countries currently suffer from decentralised and small-scale projects with limited impact, and a sectoral approach would allow for the aggregation of dispersed activities to a scale where they can become viable for a CDM (Figueres, 2005; Sterk and Wittneben, 2006).

As such, a bundled project (as opposed to its component pieces) could combine components with both low and high abatement costs, making the overall price of CERs competitive and thus feasible under CDMs. Therefore, through economies of scale, the sectoral approach could in fact create lower transaction costs that could hopefully overcome the current barrier to CDM project adoption in SDCs. Sterk and Wittneben (2006) argue that sectoral CDM activities, in addition to lowering transaction costs for projects that would otherwise not be competitive, may even pave the way for sectoral greenhouse limitation targets for developing countries by establishing data collection infrastructure.

However, it is important to recognise that while sectoral approaches may add flexibility to the current CDM and expand technology transfer opportunities for SDCs, SAs cannot eradicate the intrinsic flaws of the mechanism. First, concerns that have been raised about the host countries' sustainable development criteria, as well as the dearth of local participation opportunities in the approval process, are independent of the types of projects proposed. Nor does the sectoral approach change the CDM rationale of rewarding only a project's climate benefits, neglecting other sustainability benefits that it can provide (Sterk and Wittneben, 2006). Nonetheless, the fact that it lowers transaction costs and expands the scope of CDM to include development makes the CDM under sectoral approaches more useful as a tool for small developing countries. Most problematic, the formulation and management of aggregated small-scale projects requires that the host country have the necessary capacity to host complex projects, something that many SDCs lack. This last problem could in fact exacerbate the current geographic imbalance if projects become too complex, negating the SA benefit for SDCs. However, it seems possible that these shortcomings could be addressed by measures such as capacity building, as well as with buyers giving clear preference to projects with proven sustainable development benefits.

Using backward linkages

Given global export and import data, it is clear that small developing countries as a group are not large players in manufactured products. However, many SDCs are blessed with rich resources and are major suppliers of minerals for energy-intensive manufacturing. In many cases, SDC exports are concentrated in raw material products with little diversification, which makes them subject to fluctuations in the global market. Africa, which constitutes 37 per cent of SDCs, has been virtually the only region that has not increased its share of non-oil exports (IMF Direction of Trade Statistics); at the same time, foreign direct investment in Africa has been largely, though not exclusively, resource oriented. Looking further at the primary exports of African SDCs, the pattern of raw resource export – which dominates trade and GDP – emerges in Guinea (concentrated aluminium ore, 43.4 per cent of exports), Cameroon (crude petroleum, 43.1 per cent), Mauritania (iron ore, 39.8 per cent), Mozambique (aluminium alloy, 70.9 per cent) and Equatorial Guinea (crude petroleum, 89.6 per cent) (OECD, 2005/2006).

In resource-rich countries, current dependence on raw material extraction needs to be coupled with intelligent use of rents. Governments in SDCs can use the proceeds from raw materials to promote economic diversification and to build infrastructure that can add value to current exports. For example, the possibility of moving refining facilities closer to natural resources, i.e. moving steel production closer to iron ore mines, could mark the entry of some countries into energy-intensive industries currently discussed by sectoral approaches. Indeed, a 2007 IEA report found that some of the most efficient industries are located in developing countries, because they are built with the most up-to-date technology (IEA, 2007b). An analysis of SDC export streams, as well as the movement of value-added production to SDCs, could prove promising in seeing how much SDCs could benefit from the transfer of technology facilitated by sectoral approaches.

As such, SDCs could view emissions mitigation (and the associated transfer of energy-efficient technology) under sectoral approaches as an opportunity to upgrade export capacity from raw mineral products to finished goods such as petroleum and steel, which would increase value and revenue in the global market. Participation in a sectoral approach regime could mark the entry of resource-rich SDCs into energy-intensive industries. Furthermore, the use of best-available technology through the sectoral initiative conforms domestic production to international standards, essentially linking an individual market to the global value chain. In this light, it is necessary for interested SDCs to mark the significance of their exports when formulating CDM projects, ensuring that CDM projects reflect the broader goal of production upgrading and value-addition.

Capacity building and broader development goals

For many SDCs, the focus in climate change negotiations is not on mitigation, but rather on adaptation. While the UNFCCC does not explicitly use the word 'adaptation', it states in Article 3.3 that 'The Parties should take precautionary measures to anticipate, prevent or minimise the causes of climate change and mitigate its adverse effects'. To meet this end, the Marrakech Accords adopted at the 7th Conference of the Parties to the UNFCCC in 2001, established the need for LDCs to have more support from the international community in dealing with climate change, providing support for project financing in capacity building, adaptation, technology transfer and climate change mitigation. In light of small developing country vulnerability to climate change and its imminent consequences on those countries, there is a great need to include them in all discussions about mitigation and adaptation. Indeed, for the international community and for SDCs, it may in fact prove interesting and ultimately fruitful to think about mitigation as a form of adaptation, as SDCs could potentially use mitigation commitments to obtain energy-efficient technology and financing that could be used to build much-needed adaptive infrastructure and to facilitate broader development goals.

However, for many countries, insufficient capacity, as well as limited technical, financial and technological resources have made it difficult to begin adaptation efforts and for the implementation of multilateral environmental agreements (MEAs). MEAs provide for capacity building and technology transfer efforts to improve compliance, taking into account the special situations of developing countries and SDCs. As a result, the efforts of SDCs in sectoral approaches can be accompanied by foreign aid for capacity building, building the necessary technical skills, institutional capability and personnel, which can not only implement MEAs, but also contribute to increased foreign direct investment and development for the country in the future.

Conclusion

In most discussions about sectoral approaches, the focus continues to rest on the five large developing countries (India, China, Brazil, Mexico and South Africa) as competitiveness concerns revolve around the duality in mitigation action that will differentiate dirty products from Annex I and non-Annex I countries. However, carbon constrains and protectionist laws in developed countries will distort the market and will be unfavourable to SDCs and those large developing countries that supply intermediate inputs. Export and import data shows that SDCs are also equally important, if not sometimes more important, when compared to the large developing countries when it comes to mitigation and adaptation in the negotiations of a post-2012 climate change framework.

As a result, the post-Kyoto framework needs careful analysis to enhance participation. Sectoral approaches, while they have their flaws, offer incentive mechanisms that could be used as opportunities for SDCs to integrate into the world economy, enter energy-intensive markets for which their natural resources may offer a comparative advantage and build their capacity along broader sustainable development goals. Sectoral approaches cut across boundaries, making it more flexible to implement emissions reduction targets with broader participation and less transaction cost. In this way, while SAs may be more promising for SDCs, they are also a possible means to mitigate with less trade-distorting effects when compared to other measures such as border taxes. It only begs to question how sectoral approaches, if implemented and adopted by small developing countries, can live up to their promises of harmonising technology deployment while meeting developing country growth aspirations without hurting competitiveness.

7

Conclusions

Least developed countries, small vulnerable economies and small island developing states already face the development challenge in its full complexity. Many are struggling to achieve the Millennium Development Goals, particularly since they are poor and lack the necessary financial and technical resources to achieve their goals. The development conditions and economic resource constraints of smaller developing countries often exacerbate their economic and climate change vulnerabilities and inhibit their ability to adapt to climate change in social, technological and financial terms. Already, many are being left behind in terms of trade growth and competitiveness, and the development and maintenance of trade-related infrastructure might also become more difficult as countries struggle to cope with climate impacts and adaptation demands.

At the national level, smaller developing countries need to ensure that their trade and development policies are climate adapted. This requires measures to build up domestic productive capacity and diversification into sectors that are less vulnerable to climate change. With a small economic base, this is a steep challenge. It will also require an enabling international policy environment (within both the trade and climate regimes), to give these countries policy flexibility, capacity support, the necessary financial flows and improved technology outlooks that they need to secure trade competitiveness and climate-adapted sustainable development. Again, this is a steep challenge. LDCs, SVEs and SIDS have found it particularly difficult to influence and shape global responses to climate change in directions that minimise the negative trade impacts and secure financial and technological support. This group is not seen as 'part of the problem' when it comes to the causes of climate change. Despite being key impact-takers, it has been difficult for them to secure concrete and practical support for their own development needs within both the climate and trade spheres. The fact that these countries have small delegations and limited resources with which to engage in the negotiations compounds the problem.

This concluding section rehearses some of the specific trade and climate change topics that are of greatest interest to LDCs, SVEs and SIDS and includes policy-oriented conclusions raised during a multi-stakeholder dialogue on trade and climate change held in Mauritius in September 2008. This involved the researchers who have contributed to this publication, regional bodies of LDCs, SVEs and SIDS and private sector and government officials from Africa, the Caribbean and Pacific regions.

7.1 Energy and Technology

Small developing countries are largely net fuel importers. Many are remote islands or landlocked countries, which increases transport costs for goods and services like tourism, and makes fuel costs especially important within the economy.

LDCs, SVEs and SIDS lack access to technology in addressing their climate change concerns, and they face significant availability, cost and trade barriers in accessing the technologies they need to improve energy efficiency and develop the use of renewables. These are significant concerns with respect to competitiveness for remote economies, both in terms of cost and a desire to demonstrate clean production methods. A greater use of renewable energy within small economies would also deliver a considerable development dividend, as precious resources devoted to fuel import are diverted to other uses.

Constraints to addressing these challenges include an inadequate awareness in-country of energy-related challenges, a lack of technical capacity for energy audits and needs assessments, and financial resources, since the financial cost of undertaking assessments is often a barrier to small developing countries in addressing their policy concerns.

Innovation

Development and transfer of technology has emerged as a basic building block in the crafting of a post-2012 global regime on climate change. A range of technologies needed for mitigation and adaptation to climate change have already been identified. These include technologies needed for observation and monitoring of climate change, technologies for mitigation (e.g. energy-efficient and renewable-energy technologies; energy-efficient transportation technology; energy- and material-saving building and construction technologies; low-GHG emission technologies for agriculture and animal husbandry etc.) and technologies for adaptation (e.g. water-saving, water-capture and water-reuse technologies; agricultural biotechnology; disease- and pest-control technology; and flood, drought, sea-level rise, agricultural disasters and desertification-control technologies).

Several barriers to access to these technologies have been identified. Legal and policy measures have an important role in the transfer of technology, even as technology is largely transferred by the private sector. Trade liberalisation also has a role, but is an insufficient driver to the diffusion of the knowledge and technologies that will be required, especially in developing countries. A wide range of economic and trade-related instruments are needed to promote climate-relevant innovation and technology transfer, providing an 'enabling environment'.

In the climate change negotiations, intellectual property issues have become a bone of contention. Intellectual property rights have long been a tool to promote innovation and the dissemination of new ideas and inventions. Nevertheless, in some cases the excessive scope or level of protection of intellectual property rights in fact provides a disincentive for further research and development, as well as an obstacle to access to the protected knowledge by the broader public. Therefore, a balance will need to be achieved between patents and access to climate-related technologies. Furthermore, intellectual property is not necessarily the bottleneck for the present generation of technologies.

There will likely be no single answer to promoting the transfer of climate-related technologies. The UNFCCC has recognised that least developed countries, countries in the African region and small island developing states have 'specific needs and special situations' in regards to technology transfer. Many of the tools found useful in other contexts may be inadequate and even counterproductive for these countries, which have lower levels of development and other distinctive circumstances. For example, market-based technology transfer mechanisms such as foreign direct investment or joint ventures are ineffective in meeting the needs and demands of least developed countries.

There is increasing realisation that – both within and beyond the intellectual property system – existing innovation structures and activities can and should be enhanced, and more efforts are needed to get smaller developing countries on board.

Energy efficiency

According to the World Energy Outlook and other studies, energy efficiency improvements provide a cost-effective way to reduce energy bills and greenhouse gas emissions, lower the cost of energy services and diminish energy-import dependence. This potential is particularly important for developing countries with high energy import bills, be they LDCs or remotely located small island developing states.

However, energy-efficient end-use technologies are still relatively underutilised because of numerous market barriers. Governments are increasingly recognising the need to overcome these barriers through direct policy measures. The most widely deployed policies are mandatory minimum energy performance standards and energy labelling, both of which have been highly effective in improving equipment energy performance. In 2006, 57 countries representing 80 per cent of the world's population had energy efficiency standards and labelling programmes in place, with many more countries in the process of developing and expanding these mechanisms.

As the large majority of energy-using equipment is traded internationally, these policies have implications for the nature and cost of international trade. Different regional and national standards for testing and conformity assessment procedures, as well as import tariff regimes, add to the cost of trading energy-efficient products.

Testing and conformity assessment procedures are complex and reflect both historic and user difference across nations. Any multilateral decision to develop internationally agreed criteria for relatively energy-efficient goods, based on comparable test procedures, would need to be justified by the expectation that the net benefits of increased trade in goods outweigh the inconvenience of harmonisation. Future energy standards and labelling schemes would need to balance the need for accurate and useful data with the need to be simple, transparent and involve sufficiently low transaction costs to include small countries and players.

Trade liberalisation of environmental goods and services

Rapid diffusion of clean technologies will be key to climate change mitigation. The scale of diffusion required is unprecedented but essential, and all avenues will be needed. The most efficient mover of goods and technologies around the globe is trade.

Developed country producers and exporters have proposed, at the WTO, to single out environmentally friendly technologies, and among these, climate-friendly technologies, and ask countries to bring down tariff barriers to aid their diffusion.

Many developing countries are, however, reluctant to make such blanket commitments. They are concerned that they might, unintentionally, end up liberalising far more goods than just those with an environmental end use. They are also concerned about competition for their own small and medium-sized enterprises (SMEs) or possible future companies producing the same goods. Furthermore, as developing and LDC countries have in general higher tariffs on environmental and energy-efficient goods than OECD countries, tariff cuts have large implications not only on the market potential in developing countries, but also on government revenue from import taxes. For many developing countries, adaptation is a higher priority than undertaking mitigation measures. However, adaptation goods and technologies are diverse and diffuse and often involve low-tech local solutions and materials. For these reasons, they have not yet been a priority in discussions on EGS liberalisation.

However, the picture is rapidly changing with the emerging economies growing most quickly and becoming the new producers and exporters on the clean technology market. With regard to least developed countries, it is clear that possibilities of including products of export interest to them may largely be confined to environmentally preferable products (EPPs). However, few products, if any, may be explicitly linked to climate change mitigation objectives, except indirectly, for instance, EPPs harvested or gathered sustainably from rainforests.

Overall, trade liberalisation by itself may not be sufficient to promote the diffusion of climate-friendly goods. A whole host of complementary measures – regulatory, capacity building, financial and technology-related – will be required.

Border measures and energy-intensive sectors

For a few LDCs and SVEs, the issues of climate-related border measures will be of interest. These have been considered recently by the EU and US to address free-rider/competitiveness concerns with respect to goods produced in countries with climate mitigation measures in place, and those without. Some will also have an interest in the current consideration of sectoral agreements covering energy-intensive and heavily trade-exposed industries such as steel, pulp, paper, aluminium, cement and agrochemicals. While small developing countries do not face mandatory emissions cuts, given their high level of natural minerals exports, it could be strategic in the long-run for developing countries to adopt sectoral emissions targets that would help attract technology and finance support across a bundle of small-scale projects in an individual sector. Given that many are resource-rich, small developing countries could view emissions mitigation under sectoral approaches as an opportunity to upgrade export capacity from raw mineral products to finished goods. Participation in a sectoral approach regime could mark the entry of resource-rich small developing countries into energy-intensive industries. The use of 'best-available technologies' through the sectoral initiative conforms domestic production to international standards, essentially linking an individual market to the global value chain. Participation could be helpful to small developing countries if they use backward linkages and capacity building in keeping with broader, sustainable development goals.

7.2 Agriculture and Fisheries

Three key trade and climate change policy concerns are of particular significance in the fields of agriculture and fisheries:

- Labelling issues and food standards, given the proliferation of symbols being developed and applied by private sector retailers – which impose considerable costs on farmers and may lead to confusion amongst consumers. This is especially a concern with respect to airfreighted goods and the concept of 'food miles'.
- Investment incentives to promote:
  - Climate-friendly agriculture,
  - Food security, which is likely to be affected adversely as a result of climate change, and
  - Water resources management.
- The impact of potential future bunker fuel levies on SIDS/SVEs/LDC agricultural and fishery imports/exports, given their remoteness from markets.

Small developing countries need to engage with the private sector and NGOs on standards setting and labelling issues. Diversification strategies and appropriate policy and legal frameworks that support food security and adaptation to climate change in the agriculture and fisheries sectors will be vital.

Carbon labelling and the food miles debate

The role of voluntary carbon-labelling schemes is likely to grow in the future, providing consumers with the option of decreasing their personal carbon footprints. Given this context, the debate on food miles needs to be expanded to not just include road and sea transport, but also to look at the total carbon emissions of a product through the supply chain, using life-cycle analysis, and evaluate how to reduce emissions at each stage of the chain to achieve a carbon-neutral rating.

Overall, carbon-labelling schemes provide opportunities as well as challenges for developing countries. Any future carbon schemes need to balance the requirement for accurate and useful data with the need to be simple, transparent and involve sufficiently low transaction costs to include small countries and players. Labelling schemes provide opportunities for positive product differentiation and market opportunities. On the other hand, many producers are concerned that labelling and standards become barriers to market access. They see the rise in such 'non-tariff barriers' as potential obstacles to market entry and as a vehicle for green protectionism.

The concerns of developing countries is not being adequately heard in the development of private sector labelling schemes and there is limited room to address these through the international trade rules governing standards and technical regulations – namely the WTO Agreement on Technical Barriers to Trade and the WTO Agreement on Sanitary and Phytosanitary Standards (SPS), which bind member countries, not private organisations. However, there is now an ongoing debate in the issue of private sector standards within the WTO SPS Committee, questioning this notion.

Investment in agriculture

Global agricultural trade, and rules governing this trade, affect carbon management globally, as changes in land-use patterns have major impacts on the carbon balance. Agriculture is also the most vulnerable sector to climate change. In low-latitude regions, where most developing countries are found, even moderate temperature increases are likely to result in declining yields for the major cereal crops. This could increase the risk of hunger in many parts of the world. More frequent extreme events like floods and droughts may lower long-term yields by directly damaging crops at specific developmental stages. Heavy rainfall could precipitate soil erosion, resulting in substantial agricultural loss. Meanwhile, several studies in Africa have established a positive relationship between drought and animal death.

Both mitigation and adaptation measures can and should be taken within the agriculture sector. On the mitigation side, these include: cropland management, grazing land management and pasture improvement, restoration of degraded lands, livestock management, manure management and bioenergy. Many of these approaches provide win-win outcomes or co-benefits in terms of higher productivity, better management of natural resources and habitats, or the production of valuable by-products. Others require substantial investment at the global level, such as the development of low-emission rice varieties and livestock breeds. The options are not universally applicable and countries should evaluate and apply those best suited. In addition, possible negative effects or trade-offs of the measures should be considered.

Adaptation strategies include: farm-level practices such as planting different crop varieties, changing planting dates, crop and livestock diversification, adapting to a shorter growing season, rotating or shifting production between crops and livestock, and shifting production away from marginal lands. Macro-level adaptation strategies include the introduction of insurance and other financial mechanisms to mitigate risks, as well as temporary migration, where migrant farmers relocate from drought-affected areas to more favourable regions to farm and subsequently return to their villages when conditions improve.

The global community must support adaptation in the agriculture sector in developing countries. This also needs to be reflected in global trade negotiations. Trade justice should be promoted by ensuring that governments, particularly in poor countries, can choose the best solutions to end poverty and protect the environment. These may not always be free trade policies.

Bunker fuels

The regulation of greenhouse gas emissions from international transportation would potentially mean raising costs for moving goods and people around the globe, with implications for international trade. Developing countries situated in remote locations, such as SIDS, and developing countries with large trade exposure, would be particularly affected by higher bunker fuel prices. Certain service sectors, such as tourism, in particular would likely be affected by new emissions cuts in the international transport sector.

Current discussions on international levies on bunker fuels, or an international trading scheme in this area, need to consider this fact, and the principle of common but differentiated responsibilities taken into account. At the same time, the potential costs of climate regulation need to be seen in relation to their benefits, and other costs, such as fluctuations in energy costs unrelated to climate regulation. This matter is further considered in the following section.

7.3 Tourism and Transport

For remote or landlocked LDCs, SVEs and SIDS, transport costs are an important competitiveness concern. As high-emitting transport sectors come under pressure for action on mitigation, there will be a move towards more integrated transport systems with logistical approaches, traffic management and urban planning becoming increasingly significant.

Policy approaches to emission reductions in the aviation sector are likely to be based around carbon taxes and environmental levies. In this regard, action under the European Union's Emissions Trading Scheme (ETS) could impact negatively on long-haul flights, with particular concerns in respect to the costs that will be faced by independent and small airlines (including national carriers), and the impact on visitor and tourist numbers. With regard to the movement of fresh produce, diversion to freight may not be feasible and the extra charges may price freighted goods out of the market.

Opportunities exist in the form of marketing 'green airlines' and 'green destinations'. Some carriers have shown that awareness exists amongst airlines and there is a potential to capture opportunities. There is also a potential for premium pricing if the right business model is implemented.

Meanwhile, technological alternatives are often costly and require technological imports; there has also been under-investment in research and development in this area. However, investment in efficiency gains can be cost-neutral or yield financial and other competitiveness benefits. The harmonisation of policies could help LDCs, SVEs and SIDS to leapfrog to greener approaches in new areas. Small developing countries could also pool resources for research and development to develop indigenous capacities.

In common with aviation, an increase in carbon taxes and environmental levies in the shipping sector will increase freight costs and will impact negatively on cruise ships. Yet also like aviation, potential exists for eco-friendly marketing and premium pricing.

More generally, there is an issue related to the allocation of emissions. Within the airline and airfreight sectors, emissions might be regarded as being associated with exports from developing countries – either through the physical export of produce or, in the case of tourism, the export of services. Registration systems for shipping often involve flags of convenience in which small states have a strong interest, yet they may also be seeking special dispensations with respect to their traded goods. Ship owners and associations need to agree an approach in this area to avoid free riding.

Within shipping there is a considerable potential for better efficiency, fuels and design to lower emissions, but there has traditionally been under-investment in R&D this area. The costs of current alternatives, such as nuclear-powered ships, if very high and would result in a dependence on technology imports.

In the land transport sector, constraints to action by small developing countries include the high cost of energy-efficient vehicles and a high dependence on land transport for trade, especially for landlocked countries.

Finally, opportunities exist to switch to alternative fuels (e.g. bio-fuels and ethanol) and to create new industries in this sector. For some agriculture-dependent economies, this may be an important new avenue. Crops such as sugar could be targeted for fuel. For other states, a switch from trucking to rail could yield benefits.

7.4 Concluding Remarks

This report has provided an analysis of the trade and climate change agenda, which shows that least developed countries, small vulnerable economies and small island developing states have distinct concerns.

Not only do smaller developing countries face amongst the most severe physical impacts from climate change, but they are also especially vulnerable to these impacts, because of the structures of their economies and because such countries lack the necessary financial, technical and institutional capacities to analyse and address these concerns.

First, the economies of smaller developing countries are based largely and directly on environmental resources that will be heavily impacted by global warming. Key sectors include agriculture, fisheries and tourism. These economies will be increasingly challenged by climate impacts, which will be mediated for the most part by water – either too much of it or too little – and by related climatic impacts such as shifts in growing seasons, the loss of wildlife, heat impacts on both people and key ecosystems such as coral reefs, and changes in vector-borne diseases such as malaria. These trends point towards the need for strong international mitigation action on climate change to avoid the worst impacts on poor and vulnerable countries. It also implies the need for adaptation strategies at the national level within affected sectors, and within national economies as a whole, ensuring that LDCs, SVEs and SIDS can shift their economies in the longer term towards less vulnerable sectors.

Second, LDCs, SVEs and SIDS are particularly challenged in respect to their ability to adapt to climate change. Small countries, with a small resource base, find economic diversification particularly challenging. The group has achieved some successes through the development of service industries such as tourism and to a lesser extent niche markets like off-shore banking, but these have proved vulnerable in their own ways. Small and poor countries, with limited human resources, small domestic markets and high per-capita shared costs for government and public services, also lack the necessary technical, institutional and financial capacities to assess impacts, formulate policy responses and invest in economic transformation. Given the real challenges that smaller developing countries face with respect to trade and climate change, it is a matter of concern that their prospects and perspectives have remained obscure. Today's global warming is the result of, in good part, historic emissions of greenhouse gases from wealthy countries, and support to smaller developing countries in implementing the additional mitigation and adaptation costs they need to incur as a result of climate change need to be understood as a form of financial compensation for the damage that has been done and the burden this now places on highly vulnerable countries. Since capacity is such a vital part of the response to climate change, the development realities of these countries – their lower income and capacity – must also be full and integral part of the assistance provided to them.

The debate on trade and climate change has tended to focus on concerns that have their origins in the changing structure of the world economy and the current international framework on climate change. Aspects such as the discussion of border measures to address free-rider concerns, and debates on the trade liberalisation of environmentally friendly technologies and the role of intellectual property rights in promoting technology innovation, development and transfer, have been prominent. While some smaller developing countries share some of these concerns, they are of less direct interest to many. Instead, smaller developing countries have distinct priorities that are embedded within their own economies and which have received far less attention.

The analysis in this publication has shown that there are a number of matters that need further urgent consideration and action by national governments and at the international level. Some of these are set out below.

Actions at the national level

More specific national-level analysis of trade and climate change vulnerabilities is needed, particularly since trade forms such a key part of the economic and social well-being of smaller developing countries. This work could lead to planned (rather than just spontaneous) adaptation, both within key sectors and each national economy as a whole. It can also lead to a **climate-adapted trade policy response** – that is, domestic sustainable development policies that incorporate trade and other economic measures aimed to build up domestic productive capacity and promoting economic diversification to sectors and activities that are less vulnerable to climate change.

Appropriate tools and approaches are needed to achieve this analysis in the context of poor data availability and capacity constraints. The matter is urgent and should not wait until more conducive conditions prevail. Indeed, scenario building, multi-stakeholder dialogues and incremental analysis could help to establish a foundation for data collection and future decision-making. The sharing of experiences and findings amongst countries with similar trade and climate change challenges would be useful.

Renewable energy technologies and energy efficiency measures need particular attention, since they have the potential to deliver a significant development dividend. Low-carbon energy technologies can reduce fuel import bills and the call on hard-earned foreign exchange, and release these funds for core development purposes and adaptation. They will also help countries to maintain competitiveness as the world economy moves towards low-emission production methods, and support the objectives of smaller countries to take a stand and push for strong international action on mitigation. A great deal can be done at the national level through awareness campaigns, standards and labelling schemes, and policy measures to address cost barriers to adoption. While LDCs, SVEs and SIDS often have high tariffs on energy-efficient and environmental goods, tariff cuts in this area could result in significant losses of government revenues in countries where income tax revenues are either absent or small.

Indigenous technology capacities need to be developed in small developing countries because their technology interests – which are adaptation oriented and often more local or low-tech – may not be served well in other ways. This could also be an important route to trade diversification: the example of Barbados in developing an indigenous solar energy industry is interesting in this regard. The **pooling of technical resources** and capacity by smaller developing countries would help to address some of the capacity constraints that smaller countries face in this area.

Action on key 'reputational concerns' is another area where governments can help to protect their share of important markets by addressing issues that are of concern to consumers within those markets. This means delivering and highlighting their efforts towards sustainable approaches to fisheries, airfreighted agricultural produce and long-haul tourist destinations. The announcement of the Maldives government to become carbon neutral within a decade is one example where engagement in the issues may help to spur action at the national level and push forward international frameworks of assistance as well. Governments are well placed to support the work of industry groups to get the full picture out into the public domain, engage collectively on issues such as private labelling schemes and to promote appropriate industry developments that suit a particular country context.

Collective voice: recognising the difficulties that LDCs, SVEs and SIDS have experienced in the past in influencing the shape of global responses to the climate, trade and development challenges that they face, and that there is strength in numbers, smaller developing countries are encouraged to work effectively together to develop a strong collective voice in both in the climate change and trade negotiations to secure a strong outcome from Copenhagen summit in December 2009, and that their development agenda is accommodated within future trade agreements.

Actions at the international level

Smaller developing countries are **addressing specific concerns in current negotiations** on aspects such as climate-related levies on bunker and aviation fuel or flights, and the implications of proposals with respect to energy-intensive sectors like mining. Further work is needed to examine the impacts of proposed policies and measures on the most vulnerable member states, and to tailored responses that will unlock opportunities for these countries to leap forward in terms of technology and industry efficiency.

Major trading partners of smaller developing countries should support climate-adapted agricultural and industrial sector diversification. In the WTO, a genuine effort is needed to respond to developing country proposals for an integrated and development-oriented approach in determining the outcomes of the Doha Development Agenda. Smaller developing countries need **policy space, finance, technologies and improved capacities** to help them pursue climate-adapted economic development policies. The priorities of the negotiations must also effectively address the priorities of this highly vulnerable group.

While the onus for action in response to the trade and climate change challenges is on national governments, LDCs, SVEs and SIDS face severe capacity constraints in formulating and implementing that response. They require **a positive enabling environment** at the international level, and support that recognises and fully integrates their development realities – the lack of resources and capacity – into co-operative frameworks of assistance that are driven by national priorities.

Bibliography and References

Agroforestry Centre (2008) Avoided Deforestation with Sustainable Benefits, obtained from http://www.asb.cgiar.org/pdfwebdocs/Avoided-Deforestation-with-Sustainable-Benefits-flyer.pdf [13 June 2008].

Ahman, M (2008) 'Why International Transport Needs a Sectoral Approach'. In *Beyond Bali: Strategic Issues for the post-2012 Climate Change Regime*. Brussels: Centre for European Policy Studies.

Allison et al. (2005) Effects of climate change on the sustainability of capture and enhancement fisheries important to the poor: analysis of the vulnerability and adaptability of fisherfolk living in poverty. http://www.fmsp.org.uk/Documents/r4778j/r4778j_1.pdf [accessed 13 March 2009].

American Electric Power (AEP) (2007) Testimony of American Electric Power submitted to Senate Subcommittee on Private Sector and Consumer Solutions to Global Warming and Wildlife Protection. Washington, DC: AEP.

Andersen, Stephen, K Madhava Sarma and Kristen Taddonio (2007) *Technology Transfer for the Ozone Layer: Lessons for Climate Change*. London: Earthscan.

Anger, Niels, and Jayant Sathaye (2008) Reducing Deforestation and Trading Emissions: Economic Implications for the post-Kyoto Carbon Market. Centre for European Economic Research, obtained from ftp://ftp.zew.de/pub/zew-docs/dp/dp08016.pdf [13 July 2008].

Appleton, AE (2007) Supermarket Labels and the TBT Agreement: 'Mind the Gap'. *Business Law Brief*, Fall 2007.

Barnsley, Ingrid, and Richard Baron (2008) Possible Models for International Sectoral Approaches. Sectoral Approaches to International Climate Policy. Paris: International Energy Agency.

Baron, Richard (2006) Sectoral approaches to GHG mitigation: scenarios for integration. Paris: International Energy Agency.

Barton, J (2007) Intellectual Property and Access to Clean Energy Technologies in Developing Countries, An Analysis of Solar Photovoltaic, Biofuel and Wind Technologies. Geneva: International Centre for Trade and Sustainable Development (ICTSD).

BBC News (2008) Carbon market's value hits $64bn. 7 May 2008, obtained from http://news.bbc.co.uk/2/hi/business/7388035.stm [13 June 2008].

Benhin, JKA (2006) Climate change and South African agriculture: Impacts and adaptation options. CEEPA Discussion Paper No.21. Special Series on Climate Change and Agriculture in Africa. University of Pretoria: The Centre for Environmental Economics and Policy in Africa.

Berndes, G (2002) 'Bioenergy and water – The implications of large-scale bioenergy production for water use and supply'. *Global Environmental Change* 12: 253–271.

Bettelheim, Eric C (2008) The Voluntary Carbon Offset Market Inquiry. Ape Alliance, obtained from http://www.4apes.com/carbon/downloads/SFM-EAC-Voluntary-Carbon-Market-Inquiry.pdf [13 June 2008].

Bingley, Elsbeth (2008) Measuring and Accrediting the Carbon Footprint of Export Horticultural Produce from Kenya: Industry Attitudes and Capacity. Available at: http://ictsd.net/downloads/2008/07/paper_hasit_shah_barcelona08.pdf [accessed 15 June 2009]

Bodansky, Daniel (2007) International Sectoral Agreements in a Post-2012 Climate Framework. Working paper. Arlington: Pew Center on Global Climate Change.

Boko, M, et al. (2007) Africa, Climate Change 2007: Impacts, Adaptation and Vulnerability. Contribution of Working Group II to ML Parry et al. (eds.) Fourth Assessment Report of the Intergovernmental Panel on Climate Change. Cambridge: Cambridge University Press, pp.433–467.

Bradley, Rob, Kevin Baumert, Britt Childs, Tim Herzog and Jonathan Pershing (2007) Slicing the Pie: Sector-Based Approaches to International Climate Agreements, Issues and Options. Washington, DC: World Resources Institute.

Brander, K, G Blom, MF Borges, K Erzini, G Henderson, BR MacKenzie, H Mendes, AMP Santos and P Toresen (2003) Changes in fish distribution in the eastern North Atlantic: are we seeing a coherent response to changing temperature? *ICES Marine Science Symposia*, 219: 260–273.

Brewer, Thomas L (2003) 'The Trade Regime and the Climate Regime: Institutional Evolution and Adaptation'. *Climate Policy* 3: 329–341.

Brewer, Thomas L (2004) 'The WTO and the Kyoto Protocol: Interaction Issues'. *Climate Policy* 4: 3–12.

Briguglio, L, Bishnodat Persaud and Richard Stern (2006) Towards an Outward-Oriented Development Strategy for Small States: Issues, Opportunities and Resilience Building – A Review of the Small States Agenda Proposed in the Commonwealth/World Bank Joint Task Force Report of April 2000. 2006 Annual Meeting of the World Bank Group/International Monetary Fund, Singapore.

Brown, S, O Maseru, J Sathaye (2000) 'Project-based activities'. In R Watson, I Noble and D Verardo (eds.) *Land Use, Land-Use Change and Forestry: 'Special Report to the Intergovernmental Panel on Climate Change*. Cambridge and New York: Cambridge University Press, chapter 5.

Bueno, Ramón, Cornelia Herzfeld, Elizabeth A Stanton and Frank Ackerman (2008) *The Caribbean and Climate Change: The Costs of Inaction*. Medford/Somerville, MA: Tufts University.

Butler, Ark, and A Rhett (2008) Carbon market could fund rainforest conservation, fight climate change. Interview with Tracy Johns, Reduced Emissions from Deforestation and Forest Degradation (REDD) policy expert at Woods Hole Research Centre, obtained from http://news.mongabay.com/2008/0519-interview_johns.html [11 June 2008].

Carbon Trust (2008) Carbon Footprinting. See: http://www.carbontrust.co.uk/default.ct [accessed 15 June 2009].

Chomitz, Kenneth M (2007) Agricultural Expansion, Poverty Reduction and Environment in the Tropical Forests. World Bank, obtained from www.worldbank.org/tropicalforestreport [13 June 2008].

Claro, Edmuno, Nicolas Lucas, Mahesh Sugathan, Mario Marconini and Enrique Lendo (2007) Trade in Environmental Goods and Services and Sustainable Development: Domestic

Considerations and Strategies for WTO Negotiations. ICTSD Environmental Goods and Services Series, Policy Discussion Paper. Geneva: International Centre for Trade and Sustainable Development (ICTSD).

Climate Change Corp (2008) Pollution overboard, obtained from http://www.climatechangecorp.com/content.asp?ContentID=3513 [22 June 2008].

Cline, WR (2007) Global Warming and Agriculture: Impact Estimates by Country. Washington, D.C.: Peterson Institute.

Coalition for Rainforest Nations (2008), obtained from http://www.rainforestcoalition.org/eng/ [last accessed 16 March 2009]

Conservation International (2008) New Agreement Works To Balance Cruise Ship Tourism and Conservation, obtained from http://www.celb.org/xp/CELB/news-events/press_releases/051208.xml [22 June 2008].

Cosbey, A and R Tarasofsky (2007) Climate Change, Competitiveness and Trade. London: Chatham House.

Davis, Crystal (2008) Protecting Forests to Save the Climate: REDD Challenges and Opportunities. Earth Trends, obtained from http://earthtrends.wri.org/updates/node/303 [11 June 2008].

Danish Energy Management (2004) Benchmarking of air conditioner efficiency levels in five Asian countries. Danish Energy Management for the Australian Greenhouse Gas Office, obtained from http://www.apec-esis.org/www/UploadFile/AC_benchmarking_report_FINAL_jun04.pdf [15 June 2009].

De Fraiture, C, M Giordano and L Yongsong (2007) Biofuels and implications for agricultural water use: blue impacts of green energy. Colombo: International Water Management Institute, obtained from http://www.iwmi.cgiar.org/EWMA/files/papers/Biofuels-Charlotte.pdf

Department for International Development (2005) Growth and poverty reduction: the role of agriculture. DFID policy paper. London: DIFD.

Doornbosch, R, and R Steenblik (2007) Biofuels: Is the Cure Worse than the Disease? Roundtable on Sustainable Development. SG/SD/RT(2007)3. Paris: OECD.

Easterling, WE, et al. (2007) Food, fibre and forest products. Climate Change 2007: Impacts, Adaptation and Vulnerability. Contribution of Working Group II to ML Parry et al. (eds.) Fourth Assessment Report of the Intergovernmental Panel on Climate Change. Cambridge: Cambridge University Press, pp.273–313.

Ebeling, Johannes, and Robert Tippmann (2007) International Policy for Avoided Deforestation – Researching Synergies for Environment and Development, obtained from http://www.yale.edu/istf/files/ebeling.pdf [13 June 2008].

Ebeling, J, and M Yasué (2008a) Reducing deforestation rates 10% could generate $13B in carbon trading under REDD. Mongabay, obtained from http://news.mongabay.com/2008/0225-ebeling_redd_amazon.html [20 July 2008].

Ebeling, Johannes, and Mai Yasue (2008b) Generating carbon finance through avoided deforestation and its potential to create climatic, conservation and human development benefits. Philosophical Transactions of the Royal Society, obtained from http://journals.royalsociety.org/content/kq13503333tqvk62/fulltext.pdf [13 June 2008].

EcoSecurities (2007) Policy Brief: REDD Policy Scenarios and Carbon Markets, obtained from http://www.ecosecurities.com/Assets/10043/pubs%20-%20redd%20policy%20brief%20ecosecurities%20(background%20version)_je%20v1.pdf [13 June 2008].

Energy Information Administration (2005) World Carbon Dioxide Emissions from the Use of Fossil Fuels, International Energy Annual 2005, obtained from http://www.eia.doe.gov/iea/carbon.html [17 June 2008].

EurActive (2007) '"Carbon" import duty proposal fails to impress', 26 November 2007, obtained from www.euractiv.com [1 January 2008].

Europa (2008a) Climate Change Unit, Cruise Ship 'Costa Fortuna', obtained from http://ccu.jrc.it/costa.php [22 June 2008].

Europa (2008b) Questions and Answers on the Commission's proposal to revise the EU Emissions Trading System. MEMO/08/35. Brussels, obtained from http://europa.eu/rapid/pressReleasesAction.do?reference=MEMO/08/35&format=HTML&aged=0&language=EN&guiLanguage=en [13 June 2008].

European Commission (2008a) Europe's Climate Change Opportunity, speech by José Manuel Durão Barroso, London, 21 January 2008.

European Commission Communication (2008b) Limiting Global Climate Change to 2° Celsius: The way ahead for 2020 and beyond, obtained from www.stabilisation2005.com/programme.html [last accessed 16 March 2009].

European Parliament (2007a) Trade and Climate Change, INI/2007/2003, obtained from www.europarl.europa.eu [5 January 2008].

European Parliament (2007b) Report on trade and climate change, A6-0409/2007, obtained from www.europarl.europa.eu [19 January 2008].

European Patent Office (EPO) (2007) 'Fuelling a sustainable economy' and 'The Age of Windustry?', obtained from http://www.epo.org/topics/innovation-and-economy/emerging-technologies/ [last accessed 16 March 2009].

FAO (2006) The state of world fisheries and aquaculture 2006. Rome: Food and Agricultural Organization of the United Nations.

FAO (2007) Adaptation to climate change in agriculture, forestry and fisheries: Perspective, framework and priorities. Rome: Food and Agricultural Organization of the United Nations.

Figueres, C (2005) 'Sectoral CDM: Opening the CDM to the yet unrealized goal of sustainable development'. *International Journal of Sustainable Development Law and Policy*, Vol.2.

Figueres, C, Econergy International Corporation and Margaree Consultants Inc. (2005) Study on Programmatic CDM Project Activities: Eligibility, Methodological Requirements and Implementation. Prepared for the Carbon Finance Business of the World Bank, 29 November 2005.

Financial Times (2007) 'EU turns away from carbon tax on imports'. 25 November 2007, obtained from www.ft.com [17 December 2008].

Fischer, G, M Shah, FN Tubiello and H van Velhuizen (2005) Socio-economic and climate change impacts on agriculture: an integrated assessment 1990–2080. Philosophical Transactions of the Royal Society, 360: 2067–2083.

Foray, D (2008) Technology Transfer in the TRIPS Age: The Need for New Types of Partnerships between the Least Developed and Most Advanced Economies. Geneva: International Centre for Trade and Sustainable Development.

Fresh Produce Journal (2007) Stickers fail to take off, (Nov 2007).

Grubb, Michael, and Karsten Neuhoff (eds.) (2006a) 'Emissions trading and competitiveness'. Special issue of *Climate Policy*, Vol.6 (1).

Grubb, Michael and Karsten Neuhoff (2006b) 'Allocation and competitiveness in the EU emissions trading scheme: a policy overview.' *Climate Policy*, Vol.6 (1).

Hamwey, Robert M (2005) Environmental Goods: Where Do the Dynamic Trade Opportunities for Developing Countries Lie? Cen2Eco Working Paper. Geneva: Centre for Economic and Ecoloigcal Studies, obtained from www.cen2eco.org/C2E-Documents/Cen2eco-EG-DynGains-W.pdf [last accessed 16 March 2009].

Harrington, Lloyd, and Melissa Damnics (2001) Energy Labelling and Standards Programs throughout the World. Victoria, Australia: The National Appliance and Equipment Energy Efficiency Committee.

Hein, P (2004) 'Small island developing states: origin of the category and definition issues'. Chapter 1 in *Is a Special Treatment of Small Island Developing States Possible?* UNCTAD/LDC/2004/1.

House of Commons: Publications and Records (2008) Post-Kyoto: The International Context for Progress on Climate Change, obtained from http://www.publications.parliament.uk/pa/cm/cmenvaud.htm [13 June 2008].

Houser, Trevor, Rob Bradley, Britt Childs, Jacob Werksman and Robert Heilmayr (2008) Leveling the carbon playing field: international competition and US policy design. Washington, DC: Peterson Institute for International Economics, World Resources Institute.

Huq, Saleemul, Atiq Rahman, Mama Konate, Youba Sokona and Hannah Reid (2003) Mainstreaming Adaptation to Climate Change in Least Developed Countries (LDCS). London: International Institute for Environment and Development (IIED).

Huq, S, H Reid and L Murray (2006) Climate Change and Development Links. Gatekeeper Series 123.

The Independent (2008) 'Earth's Natural Defenses Against Climate Change "Beginning to Fail"'. 19 May 2007, obtained from http://www.countercurrents.org/mccarthy190507.htm [last accessed 16 March 2009].

ICC (2007) Sectoral Approaches: An International Chamber of Commerce Issue. Discussion Paper. Paris: International Chamber of Commerce.

Intergovernmental Panel on Climate Change (IPCC) (2000) *Methodological and Technological Issues in Technology Transfer*. Cambridge: University Press.

IPCC (2007a) Summary for Policymakers. Working Group II Climate Change 2007: Climate Change Impacts, Adaptation and Vulnerability. Geneva: IPCC. Obtained from http://www.ipcc.ch/pdf/assessment-report/ar4/syr/ar4_syr_spm.pdf [15 July 2008].

IPCC (2007b) Impacts, Adaptation and Vulnerability. Fourth Assessment Report, Working Group II Report, chapter 18: 'Inter-relationships between adaptation and mitigation', obtained from http://www.ipcc.ch/pdf/assessment-report/ar4/wg2/ar4-wg2-chapter18.pdf [18 July 2008].

International Centre for Trade and Sustainable Development (ICTSD) (2006) Linking Trade, Climate Change and Energy. ICTSD Trade and Sustainable Energy Series. Geneva: ICTSD.

International Centre for Trade and Sustainable Development (ICTSD) (2007a) Climate, Equity and Global Trade. ICTSD Trade and Sustainable Energy Series, Selected Issues Brief No.2. Geneva: ICTSD.

International Centre for Trade and Sustainable Development (ICTSD) / Gueye, Moustapha Kamal et al. (2007b) Trade, Climate Change and Global Competitiveness: Opportunities and Challenges for Sustainable Development in China and Beyond. Geneva: ICTSD.

International Centre for Trade and Sustainable Development (ICTSD) (2008) 'EU Climate Strategy: Border Measures Remain an Option'. *Bridges: Trade BioRes*, 8 (1): 25 January 2008. Geneva: ICTSD.

International Energy Agency (IEA) (2000) Energy Labels and Standards. Paris: OECD Publications.

IEA (2003) Cool Appliances: Policy Strategies for Energy-Efficient Homes. Paris: OECD Publications.

IEA (2004) World Energy Outlook. Paris: OECD Publications.

IEA (2006a) Light's Labour's Lost: Policies for Energy-Efficient Lighting. Paris: OECD Publications.

IEA (2006b) World Energy Outlook. Paris: OECD Publications.

IEA (2007a) World Energy Outlook. Paris: OECD Publications.

IEA (2007b) Tracking Industrial Energy Efficiency and CO2 Emissions. Paris: OECD Publications

International Institute for Environment and Development (IIED) (2008) Exploring the market for voluntary carbon offsets, obtained from http://www.iied.org/pubs/pdfs/15502IIED.pdf [last accessed 16 March 2009].

International Institute for Sustainable Development (IISD) (2007) A Summary Report of the Technical Workshop on Emissions from Aviation and Maritime Transport, 4–5 October 2007. Technical Workshop on Bunker Fuel Emissions Bulletin, Monday 8 October 2007.

International Roundtable on Reducing Emissions from Deforestation and Forest Degradation in Developing Countries, obtained from http://www.adpartners.org/news_brussels_summary.html [last accessed 16 March 2009].

International Trade Centre UNCTDA/WTO (2007) Airfreight Transport of Fresh Fruit Vegetables: A review of the environmental impact and policy options. Geneva: ITC.

Iturregui, Patricia, and Michael Dutschke (2005) Liberalisation of Environmental Goods and Services and Climate Change. Hamburg Institute for International Economics (HWWA) Discussion Paper 335.

Jha, Veena (2008) Environmental Priorities and Trade Policy for Environmental Goods: A Reality Check. ICTSD Trade and Environment Series Issue Paper No.7. Geneva: ICTSD.

Johns, Tracy (2006) New Mechanisms Needed to Address Deforestation. The Katoomba Group's Ecosystem Marketplace, obtained from http://ecosystemmarketplace.com/pages/article.opinion.php?component_id=4472&component_version_id=6543&language_id=12 [15 June 2009].

Knowledge Ecology International (2008) Selected Innovation Prizes and Reward Programs. KEI Research Note 2008:1.

Kopp, R, and W Pizer (2007) *Assessing US Climate Policy Options*. Washington, DC: Resources for the Future.

Kurukulasuriya, P and S Rosenthal (2003) Climate Change Series, Paper No.91. Washington, DC: World Bank.

Kurukulasuriya, P, et al. (2006) Will African Agriculture Survive Climate Change? The World Bank Economic Review; doi:10.1093/wber/lhl004. Oxford: Oxford University Press.

Lebel, T, JD Taupin and N D'Amato (1997) 'Rainfall monitoring during HAPEX-Sahel:1. General rainfall conditions and climatology'. *Journal of Hydrology*, 188–189: 74–96.

Legge, A, Orchard, J., Graffham, A., Greenhalgh, P., Kleih, U. (2006). *The production of fresh produce in Africa for export to the United Kingdom: mapping different value chains.* Natural Resources Institute. Funded by the United Kingdom Department for International Development.

L'Hôte, Y, G Mahé, B Somé and J Triboulet (2002) 'Analysis of a Sahelian annual rainfall index from 1896 to 2000: the drought continues'. *Hydrological Sciences Journal*, 47: 563–572.

Ludi, E, C Stevens, L Peskett and L Cabral (2007) Climate change and agriculture: Agricultural trade, markets and investment. London: Overseas Development Institute.

Luttrell, Cecilia, and Kate Schreckenberg and Leo Peskett (2007) Forestry Briefing: The implications of carbon financing for pro-poor community forestry. London: Overseas Development Institute, obtained from http://www.odi.org.uk/fecc/RESOURCES/briefing-papers/fb14-0712-communityforestry.pdf [13 June 2008].

Maathai, Wangari (2006) Climate Change a Pressing Issue for Africa. Green Belt Movement, obtained from http://www.greenbeltmovement.org/a.php?id=198 [12 June 2008].

MacGregor, J, and B Vorley (2006). Fair Miles? The concept of 'food miles' through a sustainable development lens. Sustainable Development Opinion. London: IIED.

Maskus, K (2003) Transfer of Technology and Technological Capacity Building. Geneva: ICTSD-UNCTAD.

Maskus, K et al. (2004) Patent Rights and International Technology Transfer through Direct Investment and Licensing. Duke University Law School, April 2003.

McCalla, AF and J Nash (eds.) (2007) Reforming Agricultural Trade for Developing Countries. Volume One: Key Issues for a Pro-Development Outcome of the Doha Round. Washington, DC: The World Bank.

McCulloch, N and M Ota (2002) Export horticulture and poverty in Kenya. Working Paper 174. Brighton: Institute for Development Studies (IDS).

McKibben, Bill (2007) 'Remember This: 350 Parts Per Million'. *Washington Post*, obtained from http://www.washingtonpost.com/wp-dyn/content/article/2007/12/27/AR2007122701942.html [13 June 2008].

McKinsey & Co. (2007) 'A Cost Curve for Greenhouse Gas Reduction'. *The McKinsey Quarterly*, February 2007.

McLeod, Elizabeth, and Rodney V Salm (2006) Managing Mangroves for Resilience to Climate Change. The World Conservation Union (IUCN) Resilience Science Group Working Paper Series No.2, obtained from http://news.bbc.co.uk/nol/shared/bsp/hi/pdfs/311006_iucn_mangroves_guide.pdf [last accessed 16 March 2008].

Meier, Alan (2001) 'Energy testing for appliances'. In Stephen Wiel and James E McMahon (eds.) *Energy-Efficiency Labels and Standards: A Guidebook for Appliances, Equipment and Lighting.* Washington, DC: Collaborative Labelling and Appliance Standards Program, pp.55–70.

Meier, Alan (2003) 'The Future of Energy Star and Other Voluntary Energy Efficiency Programmes', IEA/EET Working Paper EET/2003/02, March 2003. http://www.iea.org/textbase/Papers/2003/wpaper.pdf [15 June 2009].

Miles, Lera (2007) Reducing Emissions from Deforestation: global mechanisms, conservation and livelihoods. UNEP-WCMC, obtained from http://www.unep-wcmc.org/climate/pdf/Miles%202007%20RED%20main%20text.pdf [13 June 2008].

Mildon, R (2007) Past Developments in the World Fruits and Vegetable Sector. Presentation to Fresh 2007, Istanbul, Turkey, 6–8 June 2007.

Miller, A (2007) 'International Trade and Development'. In M Gerrard (ed.) *Global Climate Change and US Law*. Chicago: American Bar Association, chapter 8.

Mimura, N, et al. (2007) Small islands. Climate Change 2007: Impacts, Adaptation and Vulnerability. Contribution of Working Group II to ML Parry et al. (eds) Fourth Assessment Report of the Intergovernmental Panel on Climate Change. Cambridge: University Press, p.3.

Morgenstern, R (2007) 'Addressing Competitiveness Concerns in the Context of a Mandatory Policy for Reducing US Greenhouse Gas Emissions'. In Kopp, R, and W Pizer (2007) *Assessing US Climate Policy Options*. Washington, DC: Resources for the Future, Issue Brief 8.

Morgenstern, R et al. (2007) 'Competitiveness Impacts of Carbon Dioxide Pricing Policies on Manufacturing'. In Kopp, Raymond, J and William A Pizer (2007) *Assessing US Climate Policy Options*. Washington, DC: Resources for the Future.

Muller, B (2007) Food Miles or Poverty Eradication – The moral duty to eat African Strawberries at Christmas. Oxford Institute for Energy Studies.

Mytelka, Lynn (2007) Technology Transfer Issues in Environmental Goods and Services: An Illustrative Analysis of Sectors Relevant to Air-pollution and Renewable Energy. ICTSD Trade and Environment Series Issue Paper No. 6. Geneva: ICTSD, obtained from http://www.ictsd.org/pubs/ictsd_series/env/2007-04-L.Mytelka.pdf [accessed 16 March 2009].

National Board of Trade of the Government of Sweden (2004) Climate and Trade Rules – harmony or conflict? Obtained from: www.kommers.se [accessed 16 March 2009].

National Foreign Trade Council [of the US] (2007) 'WTO-Compatibility of Four Categories of US Climate Change Policy,' obtained from www.nftc.org [22 December 2007].

The New York Times (2008) 'Can Cruise Ships and Coral Coexist'. 15 January 2008, obtained from http://dotearth.blogs.nytimes.com/2008/01/15/can-cruise-ships-and-coral-coexist/ [19 June 2008].

Niasse, M (2005) Climate-induced water conflict risks in West Africa: recognizing and coping with increasing climate impacts on shared watercourses, Workshop of Human Security and Climate Change, 21–23 June 2005, Oslo, Norway.

Nicholls, RJ, et al. (2007) Ranking of the World's Cities Most Exposed to Coastal Flooding Today and in the Future. Paris: OECD.

Nordic Ecolabelling (2001) Ecolabelling of Refrigerators and Freezers – Criteria Document: 6 June 2001–17 June 2004 (Version 3.0), SIS Ecolabelling, SE-118 80 Stockholm, Sweden, obtained from http://www.svanen.nu/Default.aspx?tabName=CriteriaDetailEng&menuItemID=7082&pgr=37

Organisation for Economic Co-operation and Development (2004) Liberalising trade in 'environmental goods': some practical considerations. Document No. COM/ENV/TD(2003)34/FINAL. Paris: OECD.

OECD Development Centre and the African Development Bank. African Economic Outlook 2005/2006. Paris: OECD.

O'Sullivan, Robert, Charlotte Streck, Toby Janson-Smith, Jonathon Haskett, John O Niles, and Bernhard Schlamandinger (2006) Local and Global Benefits of Including LULUCF Credits in the EU ETS, obtained from: http://www.climatefocus.com/downloads/publications/LULUCF_EUETS_benefits.pdf [12 June 2008].

Parry, M, CA Rosenzweig, M Iglesias, M Livermore and G Fisher (2004) Effects of climate change on global food production under SRES emissions and socioeconomic scenarios. *Global Environment Change* 14 (1): 53–67.

Peskett, Leo, Ceilia Luttrell and David Brown (2006) Forestry Briefing 11: Making voluntary carbon markets work better for the poor: the case of forestry offsets. Overseas Development Institute, obtained from http://www.odi.org.uk/resources/download/23.pdf [15 June 2009].

Pricket, Glenn T (2008) International Climate Change Negotiations: Bali and the Path Toward a Post-2012 Climate Treaty. Conservation International, obtained from http://foreign.senate.gov/testimony/2008/PrickettTestimony080124p.pdf [11 June 2008].

The Rainforest Foundation (2008) Carbon Sunk: The Potential Impacts Of Avoided Deforestation Credits On Emissions Trading Mechanisms, obtained from http://www.rainforestfoundationuk.org/files/Carbon%20Sunk%20Report.pdf [15 June 2009].

Rajagopal, D and D Zilberman (2007) Review of Environmental, Economic and Policy Aspects of Biofuels. Sustainable Rural and Urban Development Division of The World Bank. Policy Research Working Paper 4341. Washington, DC: World Bank.

Raworth, K (2007) Adapting to climate change: what's needed in poor countries, and who should pay. Oxfam briefing paper 104. Oxford: Oxfam.

Richards, Michael, and Michael Jenkins (2007) Forestry Briefing 16: Potential and Challenges of Payments for Ecosystem Services from Tropical Forests. Overseas Development Institute, obtained from http://www.odi.org.uk/resources/download/560.pdf [15 June 2009]

Rigby, D and S Brown (2003) Organic Food and Global Trade: Is the marketing delivering agricultural sustainability? Centre for Agricultural, Food and Resource Economics, Manchester University. ESEE Frontiers Conference, February 2003.

Roberts, Paul (2005) *The End of Oil: On the Edge of a Perilous New World.* London: Bloomsbury Publishing plc.

The Royal Society (2007) Sustainable Biofuels: Prospects and Challenges. Royal Society policy document 01/08. London: The Royal Society.

Sachs, JD, JW McArthur, G Schmidt-Traub, C Kruk, M Bahadur, M Faye and G McCord (2004) Ending Africa's Poverty Trap. *Brookings Papers on Economic Activity*, 1, 117–240.

Sampson, Gary (1999), "WTO Rules and Climate Change: The Need for Policy Coherence", in W. Bradnee Chambers (ed.), *Global Climate Governance: A Report on the Inter-Linkages between the Kyoto Protocol and Other Multilateral Regimes*, United Nations University, Tokyo, chapter 4.

Sanghamitra, Sahu, and Gupta Neha (2008) Competition clauses in bilateral trade treaties: Analysing the issues in the context of India's future negotiating strategy. Working paper No. 24. Indian Council of Research in International Economic Relations.

Saunders, C, A Barber and G Taylor (2006) Food Miles – Comparative Energy/Emissions Performance of New Zealand's Agriculture Industry. Research Report 285. AERU, Lincoln University. New Zealand.

Schmidt, Jake, Ned Helme, Jin Lee and Mark Houdashelt (2006) Sector-Based Approach to the Post-2012 Climate Change Policy Architecture. Center for Clean Air Policy, Washington.

Slade, Tuiloma Neroni, H.E. Ambassador, Permanent Representative of Samoa to the United Nations and Chairman of the Alliance of Small Island States (AOSIS) (2000) Overview address at the Pacific Islands Climate Change Conference, Rarotonga, Cook Islands, 3–7 April 2000, entitled 'Linking Science And Climate Change Policy,' obtained from www.sidsnet.org/aosis/statements/08.html [19 June 2008].

Smith, P, et al. (2007) Agriculture. In Climate Change 2007: Mitigation. Contribution of Working Group III to B. Metz et al. (eds.) Fourth Assessment Report of the Intergovernmental Panel on Climate Change. Cambridge and New York: Cambridge University Press.

South Centre (2008) Financing the Global Climate Change Response: Suggestions for a Climate Change Fund (CCF), SC/GGDP/AN/ENV/3, May 2008. Geneva, Switzerland: South Centre.

Steenblik, R (2007) Biofuels – At What Cost? Government Support for Ethanol and Biodiesel in Selected OECD Countries: A Synthesis of Reports Addressing Subsidies for Biofuels in Australia, Canada, the European Union, Switzerland and the United States. Manitoba, Canada: International Institute for Sustainable Development.

Steenblik, Ronald, Scott Vaughan and Paul Waide (2006) Can Energy-Efficient Electrical Appliances be considered 'Environmental Goods'? OECD Trade and Environment Working Paper No. 2006–2004. Paris: OECD.

Sterk, Wolfgang, and Bettina Wittneben (2006) Enhancing the clean development mechanism through sectoral approaches: definitions, applications and ways forward. International Environmental Agreements, Springer Science and Business Media BV.

Stern, Nicholas (2006) *The Stern Review: The Economics of Climate Change*. Cambridge: University Press.

Stewart, Taimoon (2008) Interest of the Caribbean Community in Trade-Related Environmental Issues: Some Considerations. Geneva: IISD-ICTSD Southern Agenda on Trade and Environment.

Stigson, Bjorn et al. (2008) Global sectoral industry approaches to climate change: The way forward. CEPS Task Force Report. Brussels: Center for European Policy Studies.

Streck, Charlotte, and Sebastian M Scholz (2006) The role of forests in global climate change: whence we come and where we go, *International Affairs* 82 (5): 861–879.

Suppan, Steve (2008) 'Patent policy and sustainable cellulosic biofuels development'. *Biofuels News*, 19 May 2008.

Sustainable Forest Management Limited (2007a) The Voluntary Carbon Offset Market Inquiry. Submission to: The House of Commons Environmental Audit Committee. Memorandum by Sustainable Forestry Management Ltd.

Sustainable Forest Management Limited (2007b) The Importance of Forest-Based Carbon Credits for Sustainable Land Use, Biodiversity Conservation and Poverty Eradication. Submission to the European Commission for its Review on the Economics of Biodiversity Loss.

Sustainable Forest Management Limited (2008) Post-Kyoto: The international context for progress on climate change. Submission to: The House of Commons Environmental Audit Committee. Memorandum by Sustainable Forestry Management Ltd.

Swallow, Brent, et al. (2007) Opportunities for Avoided Deforestation with Sustainable Benefits: An interim report of the ASB partnership for the Tropical Forest Margin. ASB, obtained from http://www.asb.cgiar.org/PDFwebdocs/Report-on-Opportunities-for-Avoided-Deforestation-Sustainable-Benefits-web-low.pdf [13 June 2008].

The Sydney Morning Herald (2008) World May be Heating Quickly: Scientist. 7 May 2008, obtained from http://news.smh.com.au/world-may-be-heating-quickly-scientist/20080507-2bul.html [13 June 2008].

Taiyab, Nadaa (2006) Exploring the market for voluntary carbon offsets. International Institute for Environment and Development (IIED), obtained from http://www.iied.org/pubs/pdfs/15502IIED.pdf [13 June 2008].

Third World Network (2008) Brief Note on Technology, IPR and Climate Change. 23 February 2008.

UK House of Commons Environmental Audit Committee (2007) The Voluntary Carbon Offset Market. Sixth Report of Sessions 2006–07, obtained from http://www.publications.parliament.uk/pa/cm200607/cmselect/cmenvaud/331/33102.htm [13 June 2008].

United Nations (UN), General Assembly (2000) Millennium Declaration. A/RES/55/2, 18 September 2000. New York: United Nations.

UN (2002a) Report of the International Conference on Financing for Development. A/CONF.198/11, 22 March 2002. New York: United Nations.

UN (2002b) Report of the World Summit on Sustainable Development. A/CONF.199/20, 4 September 2002. New York: United Nations.

UN (2004) Report of the Secretary-General: Review of progress in the implementation of the Programme of Action for the Sustainable Development of Small Island Developing States. E/CN.17/2004/8, 11 March 2004. New York: United Nations.

UN Conference on Trade and Development (UNCTAD) (1998) The Role of Publicly Funded Research and Publicly Owned Technologies in the Transfer and Diffusion of Environmentally Sound Technologies. Background Paper No.22, Commission on Sustainable Development, Sixth Session. Geneva: UNCTAD.

UNCTAD (2000a) Bangkok Plan of Action. TD/386, 18 February 2000. Geneva: UNCTAD.

UNCTAD (2000b) Bangkok Declaration: Global Dialogue and Dynamic Engagement. TD/387, 18 February 2000. Geneva: UNCTAD.

UNCTAD (2004a) UNCTAD XI – The Spirit of Sao Paulo. TD/L.382, 25 June 2004. Geneva: UNCTAD.

UNCTAD (2004b) Sao Paulo Consensus. TD/410, 25 June 2004. Geneva: UNCTAD.

UNCTAD (2006a) Assessment of Progress in the Implementation of the Programme of Action for LDCs for the Decade 2001–2010: UNCTAD's Contribution to the Mid-Term Review, UNCTAD/LDC/2006/3. Geneva: UNCTAD.

UNCTAD (2006b) Trade and Development Report 2006: Global partnership and national policies for development. Geneva: UNCTAD.

UNCTAD (2007a) Report of the Secretary-General of UNCTAD to UNCTAD XII. Globalization for Development: Opportunities and Challenges. TD/413, 4 July 2007. Geneva: UNCTAD.

UNCTAD (2007b) The Least Developed Countries Report 2007. Geneva: UNCTAD.

UNCTAD (2007c) Trade and Development Report 2007. UNCTAD/TDR/2007. Geneva: UNCTAD.

UNCTAD (2008) The Least-Developed Countries Report 2008: Growth, Poverty and the Terms of Development Partnership. Geneva: UNCTAD.

United Nations Department of Economic and Social Affairs (UN DESA) (2006) World Economic and Social Survey 2006: Diverging Growth and Development. New York: UN DESA.

UN DESA (2007) World Economic Situation and Prospects 2007. New York: UN DESA.

UN DESA (2008) Johannesburg Plan of Implementation. Paragraph 45, obtained from http://www.environment.gov.za/nssd_2005/Web/Introduction%20and%20Background/JPOI_Response_Strategy_2003.pdf [12 June 2008].

UN Development Programme (UNDP) (2005) Human Development Report 2005: International cooperation at a crossroads – aid, trade and security in an unequal world. New York: UNDP.

UNDP (2007) Human Development Report 2007–2008. New York: UNDP.

UN Environment Programme (UNEP) (2002) Africa Environmental Outlook: Past, Present and Future Perspectives. Nairobi: UNEP, in Key Information Sheet 10/DFID, 2005 Vulnerability. Cambridge: University Press.

UNEP (2003a) Interlinkages between Biodiversity and Climate Change, obtained from http://www.cbd.int/doc/publications/cbd-ts-10.pdf [18 July 2008].

UNEP (2003b) Developing International Payments for Ecosystem Services (IPES): Avoided Deforestation, obtained from http://www.unep.ch/etb/events/pdf/AD_WGRI%202%20Side-Event%20Summary_final.pdf [15 June 2009].

UNEP (2007) Global Environment Outlook 4, State-and-Trends of the Environment: 1987–2007, ch.5, 'Biodiversity', pp.176–180.

UNEP (2008) Climate Change and the World's Water, with a Specific Focus on the Sustainable Development, Land Use and Forests. 29 January 2008, obtained from http://www.unep.org/Documents.Multilingual/Default.asp?DocumentID=527&ArticleID=5739&l=en [13 June 2008].

United Nations Framework Convention on Climate Change (UNFCCC) (2005) Climate Change: Small Island Developing States. Bonn: UNFCCC.

UNFCCC (2006a) Issues relating to reducing emissions from deforestation in developing countries and recommendations on any further process, obtained from http://unfccc.int/resource/docs/2006/sbsta/eng/misc05.pdf [13 June 2008].

UNFCCC Secretariat (2006b) Synthesis report on technology needs identified by Parties not included in Annex I to the Convention. Bonn: UNFCCC Subsidiary Body for Scientific and Technological Advice.

UNFCCC (2007a) Climate Change Impacts, Vulnerability and Adaptation in Developing Countries. Bonn: Climate Change Secretariat, UNFCCC.

UNFCCC (2007b) Vulnerability and Adaptation to Climate Change in Small Island Developing States. Bonn: UNFCCC.

United States Senate (2001) Winrock International. Senate Agriculture Committee Testimony. 29 March 2001.

Van Hauwermeiren A, G Coene, G Engelen and E Mathijs (2007) Energy life-cycle inputs in food systems: A comparison of local versus mainstream cases. *Journal of Environmental Policy and Planning* 9(1): 31–51.

Vattenfall (2007) Global Mapping of Greenhouse Gas Abatement Opportunities up to 2030, obtained from http://www.vattenfall.com [11 June 2008].

Waide, Paul (2000) Cold II: The revision of energy labelling and minimum energy efficiency standards for domestic refrigeration appliances, Contract DG-TREN SAVE N°XVII/4.1031/Z/98-269, DG-TREN, the European Commission, Brussels.

Waide, Paul, and Nathalie Bernasconi-Osterwalder, International Institute for Sustainable Development (2008) Standards, Labelling and Certification. Trade and Climate Change Seminar, 18–20 June 2008, Copenhagen, Denmark.

Williams, A (2007) Comparative study of cut roses for the British market produced in Kenya and the Netherlands. Bedford: Cranfield University.

Winjum, JK, RK Dixon and PE Schroeder (1993) 'Forest Management and Carbon Storage: An Analysis of 12 Key Forest Nations'. *Water, Air, and Soil Pollution*, 70, pp.239–57.

World Agroforestry Centre (2007) Avoided Deforestation with Sustainable Benefits: A simple way to reduce carbon emissions from deforestation and degradation. (This document is an edited version of a formal submission to the UNFCCC by the World Agroforestry Centre (ICRAF) in March 2007), obtained from http://www.asb.cgiar.org/pdfwebdocs/avoided-deforestation-with-sustainable-benefits-flyer.pdf [21 July 2008].

World Bank (2007a) International Trade and Climate Change: Economic, Legal and Institutional Perspectives. Washington, DC: World Bank.

World Bank (2007b) World Development Report 2008: Agriculture for Development. Washington, DC: World Bank.

World Bank (2008a) Country Data. Washington, DC: World Bank, obtained from http://wwwr.worldbank.org/data/countrydata/countrydata/ [31 May 2008].

World Bank (2008b) International Trade and Climate Change. Washington, DC: World Bank.

World Resources Institute (WRI) (2008) Levelling the Carbon Playing Field. Accessed at http://www.wri.org/stories/2008/05/leveling-carbon-playing-field [9 July 2008].

World Trade Organization Appellate Body (1996), Report of the Appellate Body: United States – Standards for Reformulated and Conventional Gasoline, WT/DS2/AB/R, 20 May 1996. Geneva: WTO.

World Trade Organization Appellate Body (1998) Report of the Appellate Body: United States – Import Prohibition of Certain Shrimp and Shrimp Products. WT/DS58/AB/R, 12 October 1998. Geneva: WTO.

WTO (2000) Agricultural Trade Performance by Developing Countries 1990–1998. G/AG/NG/S/6. Geneva: Committee on Agriculture, WTO.

WTO (2001) Ministerial Declaration. WT/MIN(01)/DEC/1, 20 November 2001. Geneva: WTO.

WTO Secretariat (2002) Small Economies – A Literature Review. WT/COMTD/SE/W/4, 23 July 2002. Geneva: WTO.

WTO (2002) SVEs: Issues relating to the Trade of Small Economies (Revision). WT/COMTD/SE/W/1/Rev.1, 3 May 2002. Geneva: WTO.

WTO (2004) Assistance to Address Supply-Side Constraints. WT/COMTD/LDC/W/33, 26 May 2004. Geneva: WTO.

WTO (2005) Ministerial Declaration – Annex F: Special and Differential Treatment – LDC Agreement-specific proposals. WT/MIN(05)/DEC, 22 December 2005. Geneva: WTO.

WTO (2007) Non-Tariff Measures on Products of Export Interest to the Least Developed Countries – Revision. WT/COMTD/LDC/W/39/Add.1/Rev.2, 26 February 2007. Geneva: WTO.

WTO (2008a) Market Access for Products and Services of Export Interest to Least Developed Countries. WT/.COMTD/LDC/W/41/Rev.1, 28 February 2008. Geneva: WTO.

WTO (2008b) Revised Draft Modalities for Agriculture. TN/AG/W/4/Rev.2, 19 May 2008. Geneva: WTO.

Zetterberg, L (2008) 'How to Integrate International Aviation and Shipping ino a Global Emissions Trading System'. In *Beyond Bali: Strategic Issues for the Post-2012 Climate Change Regime.* Brussels: Centre for European Policy Studies.

Index